Smart Innovation, Systems and Technologies

Volume 231

D1823291

Series Editors

Robert J. Howlett, Bournemouth University and KES International, Shoreham-by-Sea, UK

Lakhmi C. Jain, KES International, Shoreham-by-Sea, UK

The Smart Innovation, Systems and Technologies book series encompasses the topics of knowledge, intelligence, innovation and sustainability. The aim of the series is to make available a platform for the publication of books on all aspects of single and multi-disciplinary research on these themes in order to make the latest results available in a readily-accessible form. Volumes on interdisciplinary research combining two or more of these areas is particularly sought.

The series covers systems and paradigms that employ knowledge and intelligence in a broad sense. Its scope is systems having embedded knowledge and intelligence, which may be applied to the solution of world problems in industry, the environment and the community. It also focusses on the knowledge-transfer methodologies and innovation strategies employed to make this happen effectively. The combination of intelligent systems tools and a broad range of applications introduces a need for a synergy of disciplines from science, technology, business and the humanities. The series will include conference proceedings, edited collections, monographs, handbooks, reference books, and other relevant types of book in areas of science and technology where smart systems and technologies can offer innovative solutions.

High quality content is an essential feature for all book proposals accepted for the series. It is expected that editors of all accepted volumes will ensure that contributions are subjected to an appropriate level of reviewing process and adhere to KES quality principles.

Indexed by SCOPUS, EI Compendex, INSPEC, WTI Frankfurt eG, zbMATH, Japanese Science and Technology Agency (JST), SCImago, DBLP.

All books published in the series are submitted for consideration in Web of Science.

More information about this series at http://www.springer.com/series/8767

Xiaobo Qu · Lu Zhen · Robert J. Howlett ·
Lakhmi C. Jain
Editors

Smart Transportation Systems 2021

Proceedings of 4th KES-STS International
Symposium

Editors
Xiaobo Qu
Chalmers University of Technology
Göteborg, Sweden

Lu Zhen
Shanghai University
Shanghai, China

Robert J. Howlett
KES International
Shoreham-by-sea, UK

Lakhmi C. Jain
KES International
Shoreham-by-sea, UK

ISSN 2190-3018   ISSN 2190-3026 (electronic)
Smart Innovation, Systems and Technologies
ISBN 978-981-16-2326-4   ISBN 978-981-16-2324-0 (eBook)
https://doi.org/10.1007/978-981-16-2324-0

This Springer imprint is published by the registered company Springer Nature Singapore Pte Ltd.
The registered company address is: 152 Beach Road, #21-01/04 Gateway East, Singapore 189721,
Singapore

Preface

The emerging transport solutions are emerging recently which fundamentally change our urban mobility modes and associated modelling paradigm for both single-modal transport and multi-modal transport, both private transport and public transport, both land transport and maritime transport, both passenger transport and freight transport. This is mainly attributed to the research and development in telecoms, vehicular and control technologies which make it possible for individualized high-resolution data, disruptive transport modes and the coupling and consolidation of all relevant systems such as electricity grid, communications systems, transport systems, emergency management systems and others. In order to bridge the gaps among different disciplines that have all contributed to the area of urban transport, the fourth international symposium of smart transport systems (KES STS 2021) is held in June 2021, similar to last year amidst the coronavirus pandemic, to provide a communication and collaborative platform among scholars in the broad area of smart transport. Last year, we were yet to know when this will end and can only say "together we can beat this with unity". As of now, with the fast deployment of vaccine and sharp decrease of cases in all parts of the world, we are confident that we are finally seeing the light at the end of the tunnel, and we can have a real conference in 2022 to physically meet with each other and share research findings and thoughts.

For the year 2020, 17 papers were accepted in the areas of public transport, travel behaviours, maritime transport, drones, intelligent transport, traffic engineering operations, connected and automated vehicles, multi-modal transport, pandemic mitigation, driving behaviours and traffic safety. Two papers are in the area of maritime transport, discussing port shore power and seafarer change at seaports. Two papers are related to new observations using drones and Internet of things. Surprisingly, five papers (30%) are focused on public transit electrification, covering topics of optimal routing, scheduling, charging design, energy consumption prediction and pandemic influence. Three papers are in the area of traditional traffic engineering, one for weather impact, one for sustainability analysis and the last one for travel time reliability. The last five papers are focused on travel behaviour, driving behaviour, public awareness, collision warning system and cooperative control of connected and automated vehicles. These papers were rigorously peer-reviewed by at least two independent assessors and one editorial member. We establish a dialogue between assessors and authors in the progress of improving these papers. Participants and authors are mainly scholars and practitioners from Norway, Sweden, Austria, Canada, China, Australia, New Zealand and the USA.

Göteborg, Sweden	Xiaobo Qu
Shanghai, China	Lu Zhen
Shoreham-by-sea, UK	Robert J. Howlett
Shoreham-by-sea, UK	Lakhmi C. Jain

Contents

About the Editors

Xiaobo Qu is a full professor holding a chair of urban mobility systems, Chalmers University of Technology. Throughout his academic career, he has been endeavouring to practically improve transport safety, efficiency, equity and sustainability through traffic flow modelling, network optimization and most recently emerging technologies. In particular, his research has been applied to the improvement of emergency services, operations of electric vehicles and connected automated vehicles, and management of vulnerable road users. He has authored or co-authored over 90 journal articles published at top-tier journals in the area of transport engineering, and he is a recipient of many prestigious awards. His research has been supported by Australian Research Council Discovery Programme, Queensland Department of Transport and Main Roads, Sydney Trains, National Natural Science Foundation of China, Swedish Innovation Agency Vinnova and European Union. He is now an associate editor/editorial board member for *IEEE Transactions on Cybernetics*, *IEEE ITS Magazine*, *Journal of Transportation Engineering ASCE*, *Transportation Research Part A*, *Computer-Aided Civil and Infrastructure Engineering*, etc. He is a member of Academia Europaea-The Academy of Europe and a fellow of European Academy of Sciences.

Dr. Lu Zhen got his Ph.D. from Shanghai Jiao Tong University. He has ten years of teaching experience and seven years of corporate experience. He got his Ph.D. in 2008 from Shanghai Jiao Tong University. During 2008–2010, he made the postdoctoral research fellow in the National University of Singapore. His research interests lie mainly in decision support systems; knowledge management; information systems; optimization and simulation and port operations.

Dr. Robert J. Howlett is the executive chair of KES International, a non-profit organization that facilitates knowledge transfer and the dissemination of research results in areas including intelligent systems, sustainability and knowledge transfer. He is a visiting professor at Bournemouth University in the UK. His technical expertise is in the use of intelligent systems to solve industrial problems. He has been successful in applying artificial intelligence, machine learning and related technologies to sustainability and renewable energy systems; condition monitoring, diagnostic tools and systems; and automotive electronics and engine management systems. His current research work is focused on the use of smart microgrids to achieve reduced energy costs and lower carbon emissions in areas such as housing and protected horticulture.

Dr. Lakhmi C. Jain, Ph.D., M.E., B.E. (Hons) and a fellow (Engineers Australia), is with the University of Technology Sydney, Australia, and Liverpool Hope University, UK. Professor Jain serves the KES International for providing a professional community the opportunities for publications, knowledge exchange, cooperation and teaming. Involving around 5000 researchers drawn from universities and companies worldwide, KES facilitates international cooperation and generates synergy in teaching and research. KES regularly provides networking opportunities for professional community through one of the largest conferences of its kind in the area of KES.

Scheduling Shore Power Usage at Port

Yu Guo, Yiwei Wu, Wei Wang, and Shuaian Wang

Abstract In shipping industry, ships typically use their auxiliary engines to generate electrical power for lighting, ventilation and other on-board equipment and as a result emit harmful gases and particles at berth. Using shore power to power ships is a way to reduce the emissions and realize the objective of green port when ships dock at ports. But how to allocate ships to shore power is unresolved. Therefore, this paper aims to determine the allocation of shore power to maximize shore power usage and thus reduce air emissions when ship berth at the ports. An integer programming model for shore power allocation problem is proposed and it is validated by numerical experiments.

Keywords Shipping industry · Shore power · Air emissions · Green port · At berth

1 Introduction

With the development of international trade, shipping becomes more and more important, accounting for more than 90% of the total international transportation [1]. It links countries and regions, connects producers, dealers and consumers around the world, and also makes the production and consumption world-wide.

In shipping industry, there are two main types of marine transportation operations: liner transportation (traversing regular routes on fixed schedules) and tramp transportation (not having fixed schedule or port of calls but available at short notice to transport cargo from port to port). Over 95% of world shipping is powered by diesel that causes serious environment pollution by emitting harmful gas (SOx and NOx) and particles (PM10 and PM2.5) [2]. When ships dock at ports, diesel power is used to maintain on-board activities, generating emissions ten times greater than those from the ports' operation [3]. These emissions have a negative impact on port city residents, especially sensitive groups such as children and the elderly, as these

Y. Guo (✉) · Y. Wu · W. Wang · S. Wang
Department of Logistics and Maritime Studies, The Hong Kong Polytechnic University, Hung Hom, Kowloon, Hong Kong, China
e-mail: christine-yu.guo@connect.polyu.hk

© The Author(s), under exclusive license to Springer Nature Singapore Pte Ltd. 2021
X. Qu et al. (eds.), *Smart Transportation Systems 2021*, Smart Innovation,
Systems and Technologies 231, https://doi.org/10.1007/978-981-16-2324-0_1

emissions can penetrate into the lungs, leading to cardiovascular disease, lung cancer and asthma [4].

Sulphur Emission Control Area and Nitrogen Oxide Emission Control Area under the management of International Maritime organization (IMO) aim to reduce SOx and NOx emissions from shipping. Shipping companies have to take measures to reduce emissions. One potential measure is shore power, a technology that use onshore electricity rather than diesel to power the system of ships when ships dock at ports, reducing exhaust emissions at ports. However, there is an important factor related to the application of shore power: power capacity, which directly determines how many ships can be charged simultaneously. Suppose a large number of ships arrive at port at the same time, leading to a high demand for shore power. How to allocate shore power to the berthed ships equipped with shore power system under the limited shore power supply is the main issue of this paper.

2 Literature Review

IMO requires to reduce emissions and achieve green port. Shore power can meet requirements when the ships are docked at the ports. Thus, shore power has the potential to be widely used in the future. For example, China already built 616 shore power ports in 2018 and plans to build another 171 shore power ports in 2020 [1].

The investment of shore power is high. Thus, some studies of shore power have focused on economy benefits. In Safaga port, Seddiek et al. [5] has proved high-speed crafts (Alkahera) with shore power supply could save 49.03% of per year cost compared with shipborne auxiliary. Wang et al. [6] calculated the cost effectiveness of onshore power and compared the results to the cost effectiveness of using fuel. Santander et al. [7] researched the potential benefit of Spanish ports. Li [8] calculated pay-back period and analyzed related influence factors (such as service charge, power price and maintenance cost), giving suggestion that Electricity Board of China shall further reduce shore power price.

Some studies have focused on specific port(s) and discuss the actual environmental benefits. Theodoros [9] analyzed the implementation of shore power and found that the environmental benefits are significant. Chang and Wang [10] concluded that CO_2 emissions could be reduced to 57.16% by using shore power technology. The quantitative calculations used by Tseng and Pilcher [11] provided that implementing shore power in the Port of Kaohsiung would have environmental benefits in the future. Vaishnav et al. [12] showed that air quality would benefit per year if one-fourth to two-thirds of the ships that berth at the US ports used shore power supply.

Most of the current studies about shore power focus on economy and environmental benefits. However, there are very few articles on operational optimization of shore power system, such as shore power allocation, which could greatly influence total amount of shore power used and thus the amount of air emissions. Therefore, how to allocate shore power under the limited shore power supply is an urgent problem to be studies, which is the main topic of this paper.

3 Model Formulation

We consider a port equipped with several shore power sockets and providing P power (kW) to ships. Port managers plan the shore power allocation for the next planning horizon at the end of the current one with the hope to maximize shore power usage for berthed ships equipped with shore power in order to effectively reduce at-berth emission. The planning horizon is discretized into units of Δ hours (e.g., $\Delta = 0.5$). A set V of ships equipped with shore power dwell at berth in the port during a set T of time periods. Each ship $v \in V$ arrives at the port at the beginning time period a_v and leaves the port at the beginning time period b_v. We denote by p_v the power (kW) required by ship v using shore power at berth. As seafarers have to connect sockets with electrical equipment in order to use shore power, frequent connecting to and disconnecting from shore power create extra work for seafarers on board and are not conducive to equipment maintenance. Hence, we denote by u_v minimum required time of using shore power if ship v is connected to shore power in order to ensure that the ship is not constantly connected to and disconnected from shore power. Besides, different seafarers have different proficiency in operating shore power on and off. For example, some specialized seafarers may need as little as half an hour to connect a ship to shore power or disconnect a ship from shore power, while unskilled seafarers may need two or three hours to complete the operation. We use w_v to represent the time required for having the berthed ship v plug into/off the shore power.

This study proposes an integer programming (IP) model to optimally determine shore power allocation during the planning horizon in order to maximize the total amount of shore power (kW h) used by all berthed ships. Maximizing the usage amount of shore power means minimizing the amount of fuel consumed and minimizing the emissions of air pollutants and greenhouse gases. Before introducing the objective and constraints considered in this study, we list the notation used in this paper as follows.

Indices and sets:

V set of berthed ships equipped with shore power, $v = 1, 2, ..., |V|$;
T set of time periods of the planning horizon, $t = 1, 2, ..., |T|$;

Parameters:

P total power (kW) that the port can provide;
a_v time period (Δ) at the beginning of which ship v arrives at the port;
b_v time period (Δ) at the beginning of which ship v leaves the port;
p_v power (kW) required by ship v using shore power at berth;
u_v minimum required time (Δ) of using shore power if ship v is connected to shore power;
w_v time (Δ) required for having the berthed ship v plug into/off shore power;
Δ duration (h) of one time period;

Decision variables:

α_{vt} binary, set to one if and only if berthed ship v uses shore power during the ith time period, and zero otherwise;

β_v binary, set to one if and only if berthed ship v uses shore power at berth, and zero otherwise.

Based on the above definitions of indices, sets, parameters and variables, we formulate an IP model as follow:

$$[\text{M1}] \quad \text{Maximize } Z = \sum_{v \in V} \sum_{t \in T} \Delta p_v \alpha_{vt} \tag{1}$$

$$\text{s.t.} \quad \sum_{v \in V} p_v \alpha_{vt} \leq P \quad \forall t \in T \tag{2}$$

$$\sum_{t \in T} \alpha_{vt} \geq u_v \beta_v \quad \forall v \in V \tag{3}$$

$$\beta_v \geq \alpha_{vt} \quad \forall v \in V, \ t \in T \tag{4}$$

$$\sum_{t \in \{1,2,\ldots,|T|-1\}} |\alpha_{v,t+1} - \alpha_{vt}| \leq 2 \quad \forall v \in V \tag{5}$$

$$\alpha_{vt} \in \{0, 1\} \quad \forall v \in V, \ t \in T \tag{6}$$

$$\alpha_{vt} = 0 \quad \forall v \in V, \ t \in \{1, 2, \ldots, a_v + w_v - 1, b_v - w_v + 1, \ldots, T\} \tag{7}$$

$$\beta_v \in \{0, 1\} \quad \forall v \in V. \tag{8}$$

Objective function (1) maximizes the total amount of shore power used (kW h), which can be calculated by multiplying the power and shore power using time of all ships. Because the planning horizon is discretized into units of Δ hours, the time of using shore power can be calculated by $\sum_{t \in T} \Delta \alpha_{vt}$. Constraints (2) ensure that the sum of power used by all ships at any given time does not exceed the total power that the port can provide. Constraints (3) ensure that once the ship is connected to shore power, the shore power using time should not be less than the minimum required time of using shore power. Constraints (4) introduce the relationship between the variable β_v and the variable α_{vt}. Constraints (5) ensure that when ships use shore power, they need to use shore power continuously rather than connect to shore power several times. Constraints (6)–(8) define the domains of decision variables. Constraints (7) also imply that the ship cannot use shore power when connecting to and disconnecting from the shore power system.

4 Computational Experiments

In order to evaluate the proposed model, we perform several computational experiments on a PC (Intel Core i7; Memory, 16 GB). The mathematical model proposed in this study was coded in C# and implemented in CPLEX 12.5.1.

4.1 Experimental Setting

We first summarize the setting of our parameter values. The planning horizon of shore power allocation is one day (24 h). The value of Δ is set to 0.5 h. Hence, the planning horizon studied in this paper has 48 time periods. We set the values of a_v and b_v to follow the uniform distribution over the first 24 time periods and the last 24 time periods, respectively. The average value of u_v is set to 5 h (i.e., 10 time periods), which is consistent with related works [6]. The average value of w_v is set to 0.5 h (i.e., 1 time period), which is basically the same with realistic examples [13]. The average value of p_v is set to 1000 kW, which is line with Guangdong News [14]. Chu [15] shows the total power that the Suzhou port can provide is 70,800 kW, and there are three port areas in Suzhou. Therefore, we set the value of P to 23,600 kW.

4.2 Performance of the Model

A number of numerical experiments over instances with different number of berthed ships were carried out to validate the proposed model. Table 1 lists the results provided by CPLEX directly. In Table 1, 'Obj' represents the objective function values of the model solved by CPLEX directly, which are the total amount of shore power used (kW h), and 'Time' represents the CPU running time. We find that the CPU running time increases with the instance scale. Besides, although the objective function value of the model increases with the instance scale, the rate of the increase is obviously much

Table 1 Results provided by CPLEX directly

Case ID	Obj	Time (s)
Case 1 (10)	86,390	1
Case 2 (20)	208,719	3
Case 3 (30)	334,480	4
Case 4 (40)	377,524	23
Case 5 (50)	425,641	340
Case 6 (60)	440,341	1448

Note In 'Case ID', the numbers in brackets denote the numbers of berthed ships

slower from Case 3. This result is as expected because the port provides shore power to a maximum of about 23 ($23,600 \div 1000 \approx 23$) berthed ships at any point. This means that increasing the number of berthed ships does not always mean increasing the amount of shore power used because of the total power limit.

5 Conclusion

Air pollutions emitted by ships when docking at ports are of particular concern. Shore power permits berthed ships to shut down their auxiliary generators, providing a locally emission-free solution by shifting the energy generation to the shore supply. This study aims to determine the allocation of shore power to maximize the shore power usage, reducing air emissions when ship dock at ports. An integer programming (IP) model is proposed to allocate shore power during the planning horizon and this model is validated by numerical examples. Results show that the amount of shore power used does not increase with the increasing number of berthed ships because of the total power limit.

Future research work should concentrate on calculating the efficiency of shore power usage, the amount of emissions reduced at berth, the cost and benefits of allocating shore power as well as the total power demand required by the berthed ships based on ship calls information.

Acknowledgements This research is supported by the Research Grants Council of the Hong Kong Special Administrative Region, China (Project number 15201718).

References

1. Ministry of Transport of the People's Republic of China: China's port shore power construction and usage. China Shipowner's Association. Accessed 29 Oct 2020, http://www.csoa.cn/doc/17061.jsp (2020)
2. Hu, W.: The proportion of ship pollutants is increasing year by year. How can port cities deal with ship pollution. China Economic Weekly. Accessed 5 Oct 2020, http://www.qikan.com.cn (2018)
3. Habibi, M., Rehmatulla, N.: Carbon emission policies in the context of the shipping industry. Working paper, CASS Business School, London (2009)
4. Tian, L.-W., Ho, K.-F., Louie, P., Qiu, H., Pun, V., Kan, H.-D., Yu, L., Wong, T.-W.: Shipping emissions associated with increased cardiovascular hospitalizations. Atmos. Environ. **74**(3), 320–325 (2013)
5. Seddiek, I.S., Mosleh, M.A., Banawan, A.A.: Fuel saving and emissions cut through shore-side power concept for high-speed crafts at the red sea in Egypt. Marine Sci. Appl. **12**(4), 463–472 (2013)
6. Wang, H., Mao, X., Rutherford, D.: Costs and Benefits of Shore Power at the Port of Shenzhen, pp. 21–25. International Council on Clean Transportation, Washington, DC (2015)
7. Santander, A., Aspuru, I., Fernandez, P.: OPS master plan for Spanish ports project. Study of potential acoustic benefits of on-Shore power supply at berth. In: Convention Center Creta

Maris. Hersonissos. Euronoise, p. 2887. Available at http://www.euronoise2018.eu/docs/pap ers/477_Euronoise2018.pdf (2018)

8. Li, H.-B.: Analysis for shore power economy in preventing air pollution of vessels are docked at the berth. In: 2019 4th International Conference on Advances in Energy and Environment Research, pp. 2–4. EDP Sciences, Beijing (2019). https://doi.org/10.1051/e3sconf/201 911804020
9. Theodoros, P.G.: A cold ironing study on modern ports, implementation and benefits thriving for worldwide ports. Master Thesis Greece: School of Naval Architecture & Marine Engineering, National Technical University of Athens (2012)
10. Chang, C.-C., Wang, C.-M.: Evaluating the effects of green port policy: Case study of Kaohsiung harbor in Taiwan. Transp. Res. Part D: Transp. Environ. **17**(3), 185–189 (2012)
11. Tseng, P.H., Pilcher, N.: A study of the potential of shore power for the port of Kaohsiung, Taiwan: to introduce or not to introduce. Res. Transp. Bus. Manag. **17**, 83–91 (2015)
12. Vaishnav, P., Fischbeck, P.S., Morgan, M.J., Corbett, J.: Shore Power for vessels calling at U.S. ports: benefits and costs. Environ. Sci. Technol. **50**(3), 1102−1110 (2020)
13. Princess: Princess ships clear the air with shore power connections. Princess. Accessed 14 Sept 2020, https://www.princess.com/news/backgrounders_and_fact_sheets/factsheet/Pri ncess-Ships-Clear-the-Air-with-Shore-Power-Connections.html (2020)
14. Guangdong News: The largest shore power project in China southern power grid was put into operation at Nansha Port (Guangzhou). China Central Television. Accessed 18 Sept 2020, http://www.cnr.cn/gd/gdkx/20171019/t20171019_523992566.shtml?from=singlemessage (2017)
15. Chu, N.: By 2020, Suzhou will realize the full coverage of shore power for ships, and the utilization rate of shore power for berthed ships will reach 60%. Oriental Information. Accessed 15 Sept 2020, http://chuneng.bjx.com.cn/news/20181220/950549.shtml (2018)

Security Analysis Using Deep Learning in IoT and Intelligent Transport System

Gwanggil Jeon and Abdellah Chehri

Abstract Cloud computing is an embryonic and vast field. It is cost effective, flexible and customized solution provided to customers according to their needs. Due to cloud computing customers need of having their own data centers storages and services became truncated. Cloud service providers serves customers with infrastructure, storages, security, computation resources and fault rectifications. Due to the customized and flexible in nature cloud computing created many security concerns meanwhile security is the first requirement of customer. Storages, infrastructure, software and network need security measures. In this paper we deliver different defense techniques by using which cloud computing infrastructure could be made secure. It includes different protocols, encryption techniques, and hardware and software solutions for different level of security.

Keywords ITS · Security analysis · Communication model · DDoS

1 Introduction

Cloud computing is an embryonic and enormous field of computer science, which helps in achieving effectiveness and efficiency in working capacity or handling of organizations data and IT infrastructures. Cloud computing increased the capabilities of IT infrastructure due to its low-cost solutions and scalability. Information technology companies are now maintaining the cloud computing systems/ data centers. These service providers have made different data centers and providing facilities

G. Jeon (✉)
Department of Embedded Systems Engineering, Incheon National University, Yeonsu-gu, Incheon, Korea
e-mail: gjeon@inu.ac.kr

A. Chehri
University of Quebec in Chicoutimi, 555, Boul. de L'Université, Chicoutimi, QC G7H 2B1, Canada
e-mail: achehri@uqac.ca

like computational resources, storages, security, faults rectification and server maintenance etc. Big companies are now shifting their IT infrastructure to cloud technologies. On the other hand, new companies are now not building their own data centers.
On contrary, they are using cloud computing technology and focusing on their core
business instead of maintaining data centers.

Companies like Amazon, Oracle, Microsoft and IBM etc. has developed cloud
computing systems. Cloud computing uses virtualization technology in which service
providers make virtual machines on physicals servers which helps in operating any
software and services demanded by user/customer. Customers are much concern
about security and privacy of their data because all data will travel to the user using the
network from data centers and vice versa. Data can be lost accidentally, deliberately
disclosed by any employee and modified by some attacker or other tenants [1].
"Cloud computing security is the major concerns when shared resources, access
control, privacy and identity management needs" [2]. Security can be achieved or
maintained on cloud storage, cloud infrastructure, cloud software and cloud network.

"The increment in the adoption of cloud computing and the market maturity is
growing steadily because the service providers ensure the complex security level,
compliance and regulatory. In part this growth, the cloud services will deliver the
increased flexibility and cost savings" [3].

This paper will explain the cloud computing architecture, cloud computing challenges and security. Market maturity and low-cost solution are only can be achieved
until or unless security measures are taken. Trusted Third Party and cryptography
will be discussed to ensure the availability, confidentiality and integrity (CIA Triad).
Moreover, enhancement solution for cloud computing system security may also be
discussed. Internet of things (IoT) is a system of interconnected devices like vehicles,
home appliances and other things immerse with electronics, sensors, software's and
other devices can enable these objects to communicate and share information and
data between them.

The internet of things allows people and different things to be communicated any
time anywhere any place using any network any path and any service [1, 2].

The IoT allow the objects to be sensed or controlled with network infrastructure.
This new way of connectivity is going whole laptops and smart phones. It's going
towards connected cars smart homes connected wearables, smart cities and connected
healthcare basically a connected life. Recent technological advances in electronics
have enabled the deployment of all kind of small size devices of sensing, computing,
storage and power capabilities, which has led to be opportunity of utilizing any object
as a smart and communicating for this purpose of unlimited number of applications,
but at the same time that security and confidentiality requirements are met [3–5].

Every device that uses electricity gathering data and connected to the internet
what's could go wrong? The security and privacy issues surrounding are so huge.

As with the increased technology of IoT applications and devices cyber-attacks
will also be higher and more serious threads to security and privacy than ever before
[6–8].

The large amount IoT devices used in industry, military and other key areas public
and national security. E.g. on 21 Oct 2016 a multiple distributed denial of service

(DDoS) attacks systems operated by Domain name System of several websites such as GitHub, Twitter and others. This attack just executed through a botnet consisting a large number of IoT devices including printers, IP cameras, gateways and baby monitors etc. [9–11].

This study shows general survey of all the security issues along with all possibilities of IoT architectures. This paper tells us about security requirements and challenges that are usually faced in IoT implementations and security threats and expected solutions on each layer of IoT architecture to make these technologies secure and universal. According to statistics website Statista [5], the number of connected devices around the world will productively, increase from 20.35 billion in 2017 to 75.44 billion in 2025. International Data Corporation (IDC) [6] has predicted a 17.0% compound annual growth rate (CAGR) in IoT spending from $698.6 billion in 2015 to nearly $1.3 trillion in 2019, there seems to be a consensus that the impact of IoT technologies is substantial and growing. According to Gartner report in 2020 connected devices across all technologies will reach to 20.0 Billion [5, 6].

Tahir et al. provide an IC Matric base solution for IoT security the IC Matric was earlier describe in [12] and redesign for health care solution in [13]. IC Matric technology is known segregated in to health care environment has proceeded towards debugging and inscription that allow the electronic devices save and secure use. It also provides error free transmission of data between the devices. Liu et al. provide the principle of biological immune system for the security of IoT [14] this system has a role model for building IoT security. This proposed solution has five links one is security thread detection, danger computation, security defense strategy formulation, security responses and security defense. All these links are related with the data of IoT security.

Zhou and Chao provide solution of security critical traffic management scheme. The novel media aware traffic security architecture (MSTA) meat the security information requirement for multimedia communication and services in the IoT environment [15].

The insert solution is used for the existing architecture multimedia security [16]. In this session the author proposed the solution and challenges for multimedia security in wide networks.

Lassa dos Santos et al. introduced the architecture to enable devices that use data gram transport layer security (DTLS) with authentication to communicate with internet devices [17]. This security architecture for IoT is based on third party device which is called internet of things security support provider (IoTSSP) and to main functions for brooder router: (ii) the optimal hand shaking delegation (ii) the transfer of section.

Kothmayr et al. proposed the use of hardware to enable security in devices [18]. The main thought of author to use the trusted platform modules added in each device. Also, this architecture provides message integrity, confidentiality and authentication and un expensive energy solution, end to end latency and memory overhead, that's make it suitable security solution for IoT [19].

The composition of the paper will be like this, Sect. 2 will describe the cloud computing architecture and then Sect. 3 will be comprised of cloud computing security solutions. Section 4 will cover the conclusion of this paper.

2 Cloud Computing Architecture

Security of Perception Layer Supplies, for example, RFID peruses, sensors, portals, GPS and different gadgets require to be secured proficiently. OWASP has recognized poor physical security in the best 10 IoT vulnerabilities. The initial step is to guarantee that lone approved individuals can approach touchy information created by physical items that is the reason a physical personality and access administration strategy should be characterized. Validation and approval necessities from IoT are fulfilled in this comparable form.

Information accumulation is an imperative issue for this layer.

The second heading is picture information gathering, to utilize security in pictures as picture pressure, and CRC. Cryptographic preparing is one of the principle errands in security components for sensor information on IoT. These activities that are regularly utilized as a part of request to ensure security of information incorporate encryption and unscrambling, key and hash age, and sign and confirm hashes.

Risk Assessment is an essential of IoT security which decides the degree of the potential danger and the hazard related with an IoT framework. The yield of this procedure recognizes suitable controls for decreasing or wiping out hazard amid the hazard alleviation process. Various associations have created rules for directing danger evaluation, for example, the U.S. National Institute of Standards and Technology (NIST). The International Standards Organization (ISO) and the International Electro technical Commission (IEC).

The wired security sub-layer is worried about devices, which speak with different gadgets on the IoT framework utilizing wired channels. Basic security strategies are connected in wired sort systems are firewalls and Intrusion Prevention System (IPS). On the off chance that the system has firewall or IPS, it can assess arrange packet profoundly that are foreordained towards the goal. In any case, existing IoT has no capacity as far as bundle investigation and packet sifting. There is a progressing research on this issue where security analysts endeavor to outline a low asset hungry firewall for IoT to give the capacity of packet review.

Security of Support and Application Layers contains two sub-layers. In one sublayer, there are local applications and related middleware functions which should be secured with various techniques. For example, intelligent transportation systems can use encryption techniques, while smart home/smart metering systems uses steganography techniques. The second sublayer corresponds to national applications and their security systems, ensuring that sent and received data are secure. Therefore, various security techniques are applied in these systems based on the scope of each system such as authentication, authorization, access control list, selective disclosure, intrusion detection, firewall, and antivirus.

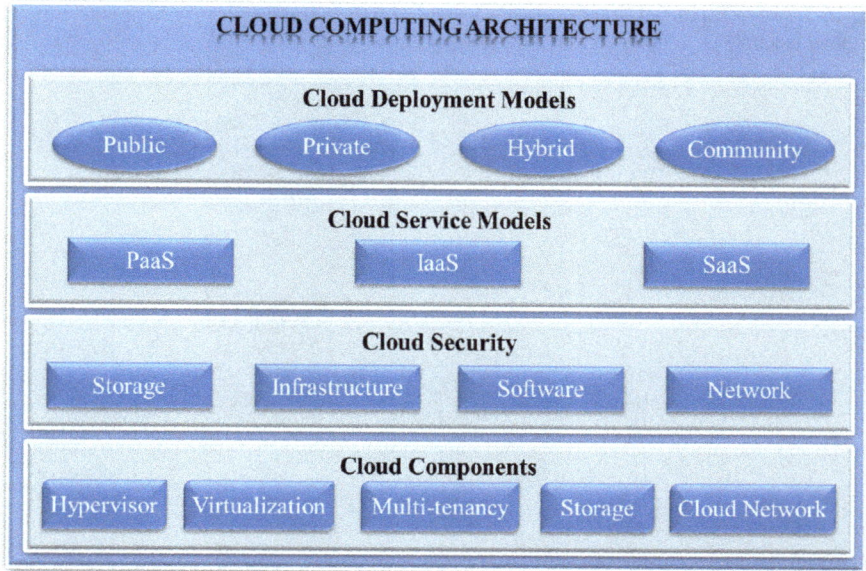

Fig. 1 Education cloud computing architecture

According to NIST, Cloud computing "is a model for allowing ubiquitous, convenient, and on-demand network access to a number of configured computing resources (e.g. networks, server, storage, application, and services) that can be rapidly provisioned and released with minimal management effort or service provider interaction". Cloud computing comprises of deployment models, service models, security and some basic components. Cloud computing architecture is presented in Fig. 1.

2.1 Cloud Deployment Model

Cloud technology can be implemented in four models i.e. Public, Private, Hybrid and Community model. They can be elaborated by the following Fig. 2.

Public Cloud: A public cloud represents the cloud hosting and owned by the service provider whereby the client and resource provider have service level agreement". Public cloud services are provided by Sun, Microsoft, IBM, Google, VMware and Amazon. That platform provides a generalized computing environment which is publically and easily available for public to use. This type is less secure than other types of cloud.

Private Cloud: Private cloud is used solely by one business, organization or one customer that may be operated by a third party or by organization itself. Private

Fig. 2 Deployment model
of cloud technology

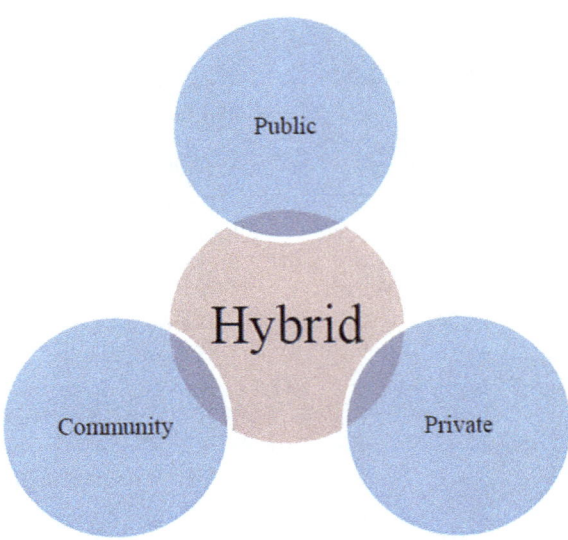

cloud provides better security at larger cost. "The St. Andrews Cloud Computing Co-laboratory and Concur Technologies are illustration associations that have a private cloud".

Community Cloud: Community cloud is the shared infrastructure by an organization with others having the same requirements or interests like same security polices, task and agreements. This type of cloud may be maintained, controlled, operated and handle by third party. Siemens provides the facility if community cloud. "The cloud infrastructure of community cloud is shared and owned by different organizations such as research groups, together with work of companies and government organizations".

Hybrid Cloud: Hybrid cloud is designed by the fusion of two or more deployment models of cloud which may be private, public or community. Above mentioned Fig. 2 also shows the overlapped area with other deployment models. In the hybrid cloud could be freely overseen yet applications and information would be permitted to move over the cloud. This is the much secure, customize and flexible type of cloud.

2.2 Cloud Services Model

Cloud architecture provides set of services/set of resources for clients according to the user requirements. There are three types of services emerged till now i.e. IaaS, Paas, and SaaS.

1. **Infrastructure as a Service (IaaS)**: Service provider of cloud offers its infrastructure with the provision of storage, physical servers, network, processing and

computational resources to the customers. Customer can use infrastructure for installation and running their own applications. Amazon Web Services are the noticeable example which uses IaaS model.

2. **Platform as a Service (PaaS)**: "Provides the capability to deploy onto the cloud infrastructure, consumer created applications, produced using set of programming languages and tools that are supported by the PaaS provider." Examples of PaaS are Microsoft Azure and Google App engine.

3. **Software as a Service (SaaS)**: Cloud provider offers its customers, applications running and installed on cloud computing infrastructure using the web browser. "The consumers do not have control or figure out how to the underlying framework including system, servers, network, operating systems, storage, or even individual application capacities, with the conceivable exemption of constrained client particular application setup settings".

2.3 Cloud Security

1. **Storage Security**: End user requires its data should be secure from attacks of third party and employees of service provider of cloud. "Privacy of the data should be maintained and remain available if the cloud provider is not available. Storage security concerns about data sanitization, cryptography, data-Remanence, data leakage, snooping of data availability and malware".

2. **Infrastructure Security**: Infrastructure of cloud may be the combination of physical and virtual infrastructures, which may be trusted by each party. "The attestation of the third party is not enough for the critical business process. It's absolutely essential for the organization to be able to verify business requirements that the underlying infrastructure is secure".

3. **Software Security**: Protection of applications or software running on cloud is also an important requirement of clients from internal and external threats. Cloud software will be developed under standard SDLC in which measures are taken against risks and threats. "It is important to define the security process and policies about the software that enables the business instead of introducing other risk and it poses challenges for the customers and the cloud provider. Software security can be handled or defeat by implementing bugs, design flaws, buffer overflow, error handling agreements".

4. **Network Security**: Cloud service provider is responsible to provide the security to its network that the data should move to and from client on a secure channel. There is another responsibility of service provider that only valid traffic can use its network and block malicious traffic [20–23].

3   Cloud Computing Security Solutions

According to the analysis of work done in the field of cloud computing, we found the following security solutions also transcribed in Table 1.

3.1   Storage Security

- AES: AES encryption is the fastest method that has the flexibility and scalability and it is easily implemented. On the other hand, the required memory for AES algorithm is less than the Blowfish algorithm. AES algorithm has a very high security level because the 128, 192 or 256-bit key are used in this algorithm. It shows Enhancement of Cloud Computing Security with Secure Data Storage using AES and resistance against a variety of attacks such as square attack, key attack, key recovery attack and differential attack. Therefore, AES algorithm is a highly secure encryption method" [8].
- RRNS: Using the RRNS (Redundant Residue Number System), this approach consists of splitting a file in different residue-segments, encrypt and upload them in different providers. The advantage of such an approach is twofold. A user can retrieve his/her file even if a provider is not temporarily or permanently available. On the other hand, providers cannot access the files stored within them [9].
- RDIC: RCID is an Identity-based remote data integrity check protocol which provides "completeness, security against a malicious server (soundness), and privacy against the TPA (perfect data privacy) to the cloud storage" [10].
- TPA: or Third-Party Auditor is an expertise and authorized client and used for auditing purpose. It helps to increase data storage correctness in cloud and used for data verification. Third party auditor detects data modification using hash code" [8].

3.2   Infrastructure Security

- SDN: SDN gives another and dynamic system design for cloud computing, the great feature of SDN make it simpler to identify and respond DDoS assaults in cloud computing. SDN controllers recognize unusual traffic, screen malicious

Table 1 Cloud computing security analysis

Cloud Security	Storage	Infrastructure	Software	Network
Defense techniques	• AES [8] • RRNS [9] • RDIC [10] • TPA [8]	• SDN [11] • Zero-day Threat detection [12]	• Traffic monitoring [13] • Cloud resilience managers [14]	• IDS [15] • IPS [15]

packet, or authenticate source IP address. "It is in favor of the community to study how to make full use of SDN's advantages to defeat DDoS attacks and how to prevent SDN itself becoming a victim of DDoS attacks in cloud computing environments" [11].

- Zero Day threat Detection: This treat should be the motive of security protection team. Threats rises gradually in reference [12] author said "To build an accurate protection system on the basis of monitoring cloud behavior we must carefully select the critical objects in the cloud that are targets for hackers and can lead to useful monitoring results".

3.3 Software Security

- Traffic Monitoring: Traffic monitoring of the network should be done which provides network forensics analysis, anomaly detection and application identification other than network management tasks. Sometimes IPSs don't have access of end-user devices. They work on network layer such as routers and switches to provide indicators. In regard, they can provide counter solutions [13].
- CRMs: CRMs stand for Cloud Resilience Managers. "The detection system should be capable of gathering and analysing data at the network component level through the deployment of network CRMs" [14].

3.4 Network Security

- IDS: Intrusion Detection System (IDS) examines the packet header and payload to compare it with any anomalies found in comparison with the normal traffic. This is in contrast to a firewall which filters the network traffic by examining the packet headers flowing through the network ports. For anomalous traffic, an IDS attempts to identify the pattern against common threats, and alerts the network administrator" [15].
- IPS: Intrusion Prevention System (IPS) works just like an IDS, however it may also reject the packets or terminate the connection. Since the backbone of a cloud-based platform is usually a high-speed network, it must be protected by a fully auto mated intrusion detection/prevention system" [15].

4 Conclusion

In this paper we discussed different security solution for cloud computing at different levels. At storage and network level data needs a great encryption and authentication. AES and RDIC can be implemented. However, at application level different network traffic monitoring, Cloud Resilience Managers and IDS/IPS can be installed for blocking unauthorized traffic and finding anomalies and malwares.

These methodologies may achieve the security demand of customers. Internet has made our life easier and efficient, with the help of internet we can communicate with each other at a large level. Through internet we can made social relationship in a professional way. IoT is required to incorporate propelled advances level of correspondence, organizing, distributed computing, detecting and activation, and make ready for important applications in a collection of regions, which will influence numerous parts of individuals' lives and achieve numerous accommodations. By the by, given the huge number of associated gadgets that are conceivably defenseless, exceedingly remarkable dangers rise around the issues of security, protection, and administration in IoT. This study centers around the security issues and difficulties of IoT, presents a review of exceptional IoT security arrangements, and exhibits a portion of the open difficulties in this field. Because of page constraints we are not ready to display the subtle elements of the considerable number of points of interest, however the expanded form of this paper will cover nitty gritty clarification of each research issue and future research directions.

References

1. Mushtaq, M.F., et al.: Cloud computing environment and security challenges: a review. Int. J. Adv. Comput. Sci. Appl. **8**(10), 183–195 (2017)
2. Gou, Z., Yamaguchi, S., Gupta, B.: Analysis of various security issues and challenges in cloud computing environment: a survey. In: Identity Theft: Breakthroughs in Research and Practice. 2017, IGI Global. pp. 221–247
3. Wilson, P.: Positive perspectives on cloud security. Inf. Secur. Tech. Rep. **16**(3–4), 97–101 (2011)
4. Mell, P., Grance, T.: The NIST definition of cloud computing (2011)
5. Modi, C., et al.: A survey on security issues and solutions at different layers of Cloud computing. J. Supercomput. **63**(2), 561–592 (2013)
6. Puthal, D., et al.: Cloud computing features, issues, and challenges: a big picture. In: 2015 International Conference on Computational Intelligence and Networks (CINE), 2015
7. Singh, S., Jeong, Y.-S., Park, J.H.: A survey on cloud computing security: issues, threats, and solutions. J. Netw. Comput. Appl. **75**, 200–222 (2016)
8. Shimbre, N., Deshpande, P.: Enhancing distributed data storage security for cloud computing using TPA and AES algorithm. In: Computing Communication Control and Automation (ICCUBEA), 2015
9. Celesti, A., et al.: Adding long-term availability, obfuscation, and encryption to multi-cloud storage systems. J. Netw. Comput. Appl. **59**, 208–218 (2016)
10. Yu, Y., et al.: Identity-based remote data integrity checking with perfect data privacy preserving for cloud storage. IEEE Trans. Inf. Forensics Secur. **12**(4), 767–778 (2017)
11. Yan, Q., et al.: Software-defined networking (SDN) and distributed denial of service (DDoS) attacks in cloud computing environments: a survey, some research issues, and challenges. IEEE Commun. Surv. Tutor. **18**(1), 602–622 (2016)
12. Ibrahim, A.S., Hamlyn-Harris, J., Grundy, J.: Emerging security challenges of cloud virtual infrastructure. arXiv preprint arXiv:1612.09059 (2016)
13. Ahmad, I., et al.: Security in software defined networks: a survey. IEEE Commun. Surv. Tutor. **17**(4), 2317–2346 (2015)

14. Watson, M.R., et al.: Malware detection in cloud computing infrastructures. IEEE Trans. Dependable Secure Comput. **13**(2), 192–205 (2016)
15. Khan, M.A.: A survey of security issues for cloud computing. J. Netw. Comput. Appl. **71**, 11–29 (2016)
16. Zhou, L., Chao, H.C.: Multimedia traffic security architecture for the internet of things. IEEE Network **25**(3), 35–40 (2011)
17. Lassa dos Santos, G., Guimarães, V.T., da Cunha Rodrigues, G., Granville, L.Z., Tarouco, L.M.R.: A DTLS-based security architecture for the Internet of Things. In: 2015 IEEE Symposium on Computers and Communication (ISCC), Larnaca, 2015, pp. 809–815
18. Kothmayr, T., Schmitt, C., Hu, W., Brünig, M., Carle, G.: DTLS based security and two-way authentication for the Internet of Things. Ad Hoc Networks **11**(8), 2710–2723 (2013)
19. Leo, M., Battisti, F., Carli, M., Neri, A.: A federated architecture approach for Internet of Things security. In: 2014 Euro Med Telco Conference (EMTC), Naples, 2014, pp. 1–5
20. Lazrag, H., Chehri, A., Saadane, R., Rahmani, M.D.: A blockchain-based approach for optimal and secure routing in wireless sensor networks and IoT. In: 2019 15th International Conference on Signal-Image Technology & Internet-Based Systems (SITIS), Sorrento, Italy, 2019, pp. 411–415. https://doi.org/10.1109/SITIS.2019.00072
21. Lazrag, H., Chehri, A., Saadane, R., Rahmani, M.D.: Efficient and secure routing protocol based on Blockchain approach for wireless sensor networks. Concurr. Comput. Pract. Exper. e6144 (2020). https://doi.org/10.1002/cpe.6144
22. Chehri, A., Mouftah, H.: Localization for vehicular ad hoc network and autonomous vehicles, are we done yet? In: Connected and Autonomous Vehicles in Smart Cities. CRC Press, Taylor & Francis Group (2020)
23. Chehri, A., Mouftah, H.: An empirical link-quality analysis for wireless sensor networks. In: 2012 International Conference on Computing, Networking and Communications (ICNC), Maui, HI, pp. 164–169 (2012). https://doi.org/10.1109/ICCNC.2012.6167403

Drone-Based Image Processing for Construction Site Safety, Transportation, and Progress Management

Wen Yi and Xiaobo Qu

Abstract Construction plays a pivotal role in securing prosperity in many countries through the deployment of new infrastructure, as the need to deliver higher-quality products more efficiently and safely is stronger than ever. Current construction management primarily relies on human assessment due to lack of automated project data collection, communication or distribution of important information, and is often reactive to accidents, inefficiencies, and progress delays. This study will tackle these problems by developing a methodological framework on automated tracking, inspection, and monitoring to increase accessibility, efficiency and safety to improve infrastructure project delivery, which integrates cost-effective unmanned aerial vehicles (UAVs, or drones) data collection, real-time analysis and segmentation of UAV visual data, and proactive dynamic planning.

1 Introduction

Construction plays a pivotal role in securing prosperity in many countries through the deployment of new infrastructure, as the need to deliver higher-quality products more efficiently and safely is stronger than ever. Current construction management primarily relies on human assessment due to lack of automated project data collection, communication or distribution of important information, and is often reactive to accidents, inefficiencies, and progress delays. This study will tackle these problems by developing a methodological framework on automated tracking, inspection, and monitoring to increase accessibility, efficiency and safety to improve infrastructure project delivery, which integrates cost-effective unmanned aerial vehicles (UAVs, or drones) data collection, real-time analysis and segmentation of UAV visual data, and proactive dynamic planning.

W. Yi (✉)
School of Built Environment, College of Sciences, Massey University, Auckland, New Zealand

X. Qu
Department of Architecture and Civil Engineering, Chalmers University of Technology, Gothenburg, Sweden
e-mail: xiaobo@chalmers.se

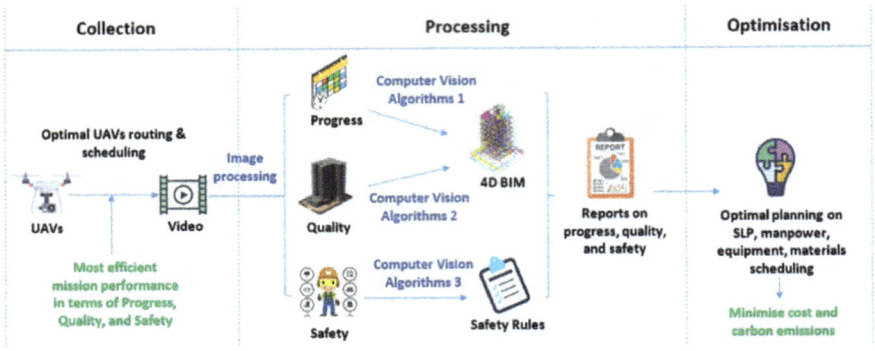

Fig. 1 The methodological framework

UAVs have been applied in various kinds of infrastructure projects (e.g., airports, railways, bridges, roads) [1, 2] and adopted as a tool for construction progress monitoring [2], safety inspection [3], earthwork surveying [4], and energy audit [5]. These applications, however, were restricted to UAV-based visual data collection methods, overlooking the need for heavy manual post-flight data processing, therefore failing to provide tangible and proactive solutions to the issues related to human intervention. While UAV-based data collection is beneficial, disruptive transformation will only be achieved through full automation of processes. Previous attempts were successful in providing proof-of-concept UAV applications aided by imaging processing algorithms for semi-autonomous and asynchronous inspection of existing infrastructure [6, 7]. However, fully automated UAV-assisted monitoring of ongoing infrastructure projects (based on synchronous data-processing) poses several challenges and has yet to be achieved [8]. Computer vision and image processing algorithms provide an opportunity to overcome the need for manual post-flight data processing and analysis.

This paper will develop a methodological framework that aims to transform the construction management of infrastructure projects by providing a fully automated platform that requires no human intervention to achieve real-time continuous monitoring and control of construction processes, without human intervention, to boost the construction sector's productivity. The framework is shown in Fig. 1. We will focus on three aspects: construction safety, transportation in construction sites, and construction progress management.

2 Construction Safety

The construction industry is full of hazards due to the heights of objects and workers and due to moving machines and construction materials. To protect construction workers from being harmed by the hazards, they are mandated to wear a number of personal protective equipment (PPE), for example, safety helmets, spectacles, headphones, gloves, and footwear. However, wearing these PPEs is usually uncomfortable for workers. For instance, it is hot to wear helmets and gloves in the summer months and workers cannot chat freely with co-workers when wearing headphones. As a result, research shows that a number of construction workers violate the safety rules regarding wearing PPEs.

To address the problem, in our methodological framework, one or several drones will be deployed, depending on the size of the construction site, to monitor the workers in the site. The drone will fly along a path that covers all the workers in the site. Of course, a worker (or a group of workers) may be covered multiple times, depending on their historical records of violations and on their job natures. Along the path, the drone will take a number of photos or a video. If there is 4G, 5G, or Wifi data connection, then the photos and videos can be transmitted to a central server in real time. Otherwise, the photos or video can be manually copied to a central server when the drone returns to its depot.

Since there will be a large number (e.g., over 1000) of photos or a long video (e.g., several hours) to process every day, it will be too time-consuming and thereby impossible to identify the workers who violate the safety rules in a manual manner. Given the advancement of image processing techniques using deep learning, in our methodological framework, image processing techniques will be employed to achieve this target efficiently and at a low cost. Specially, the image processing will consist of several tasks: (i) Identify the workers in each image; (ii) Identify the PPEs in each image; and (iii) Associate each PPE to a worker. Note that these three tasks are generally standard ones in image processing. What we need to do is to manually label several thousand images and use these images to train a deep learning model. The deep learning model can be trained using transfer learning method, that is, the structure and many parameters in the model are set to be the same as the ones in a well-trained model for other problems (e.g., for identifying whether an image is the photo of a cat) and only a small subset of parameters is trained using our data set.

Once the image processing algorithm alarms that in a particular image, a worker has a high chance of violating the safety rules, the site manager will manually check the image and draw a conclusion. If yes, the manager should immediately call the worker and ask him/her to correct the problem.

3 Transportation in Construction Sites

Construction materials and construction waste must be transported on a construction site. Using drone-based image processing techniques, we can better management the transportation of materials and waste.

3.1 Automatic Identification of Transportation Requirement

The construction materials (e.g., sand) are stored on site. Due to limited space in the site, once the amount of material is low, a new batch (usually a new truckload) of material should be ordered. However, the consumption of construction materials is uncertain and when to place a new order is often decided based on experience or on physical inspection of the site. Using drone-based image processing techniques, the amount of remaining material can be automatically calculated. Consequently, the construction manager can be alerted when the amount of material is low.

Similarly, due to limited site space, the construction waste should be transported to city dumping sites many times during the construction process. Knowing the quantity of waste is important for manger to decide when to call a truck to transport away the waste. Drone-based image processing techniques can achieve this target.

3.2 Identification of Inefficient or Unsafe Route of Transportation Machines

The transportation machines in a construction site, e.g., a tractor, travels on the site. The drivers of the transportation machines may not be familiar with the layout of the site and hence drives along a path that is not the shortest or that poses hazards to others or the driver him/herself. Using drone-based image processing techniques, we can identify the path used by the driver and hence alert the site manager if the path is not the shortest or poses safety concerns.

3.3 Congestion Management on a Construction Site

When a large number of transportation machines work in a construction site, the site can be very congested. Because of the difference in user equilibrium and system optimum traffic assignment, it can be expected that by shifting the paths of some transportation machines, the overall transportation time of the machines can be reduced. Using drone-based image processing techniques, we can identify the path and travel time of each transportation machine and then use nonlinear programming [9–13] to plan the paths of the machines, so that the total transportation time is minimized.

4 Construction Progress Management

Construction progress is an important factor that determines the supply of construction materials, the scheduling of manpower, and interference to make sure the project is completed on time. The construction progress is complex for a large project because many activities are carried out simultaneously. For instance, the installation of walls is done 50%, but the installation of the windows is done 60%, what is the overall progress of the project?

Our methodological framework aims to address the problem in this way. First, the construction project will be divided into a number of tasks, e.g., installation of walls, installation of the windows. Then, drones will be used to fly over the site and take photos/a video. The images will be processed to identify the progress of each task. Using information from building information modeling (BIM) and historical project records, we will calculate an overall progress metric. We can calculate multiple progress metrics, for instance, the progress of building 1, or the progress of wall installation. These metrics can be used by site managers to better manage the site.

5 Future Research Opportunities

The above-mentioned methodological framework should be implemented while considering all the details in the future. For example, the selection of drones, and selection of image processing algorithms, the integration of data and construction domain knowledge, and the use of data in optimization models [14]. The full implementation of the framework can bring in significant cost savings, efficiency improvement, and carbon emission reduction.

References

1. Xia, J., Wang, K., Wang, S.: Drone scheduling to monitor vessels in emission control areas. Transp. Res. Part B: Methodol. **119**, 174–196 (2019)
2. Lin, J.J., Han, K.K., Golparvar-Fard, M.: A framework for model-driven acquisition and analytics of visual data using UAVs for automated construction progress monitoring. In: 2015 International Workshop on Computing in Civil Engineering 2015, pp. 156–164 (2015). https://doi.org/10.1061/9780784479247.020
3. Irizarry, J., Gheisari, M., Walker, B.N.: Usability assessment of drone technology as safety inspection tools. J. Inf. Technol. Constr. **17**(12), 194–212 (2012)
4. Rakha, T., Gorodetsky, A.: Review of unmanned aerial system (UAS) applications in the built environment: Towards automated building inspection procedures using drones. Autom. Constr. **93**, 252–264 (2018)
5. Siebert, S., Teizer, J.: Mobile 3D mapping for surveying earthwork projects using an Unmanned Aerial Vehicle (UAV) system. Autom. Constr. **41**, 1–14 (2014)
6. Rakha, T., Liberty, A., Gorodetsky, A., Kakillioglu, B., Velipasalar, S.: Heat mapping drones: an autonomous computer-vision-based procedure for building envelope inspection using unmanned aerial systems (UAS). Technology|Architecture+Design **2**(1), 30–44 (2018)
7. Ersoz, A.B., Pekcan, O., Teke, T.: Unmanned aerial vehicle based pavement crack identification method integrated with geographic information systems. In: Transportation Research Board 96th Annual Meeting, Transportation Research Board (2017). https://trid.trb.org/view/1438958. Accessed 4 Nov 2020

8. Li, M., Zhen, L., Wang, S., Lv, W., Qu, X.: Unmanned aerial vehicle scheduling problem for traffic monitoring. Comput. Ind. Eng. **122**, 15–23 (2018)
9. Xu, M., Yang, H., Wang, S.: Mitigate the range anxiety: Siting battery charging stations for electric vehicle drivers. Transp. Res. Part C: Emerg. Technol. **114**, 164–188 (2020)
10. Cheng, Q., Wang, S., Liu, Z., Yuan, Y.: Surrogate-based simulation optimization approach for day-to-day dynamics model calibration with real data. Transp. Res. Part C: Emerg. Technol. **105**, 422–438 (2019)
11. Liu, Z., Wang, S., Zhou, B., Cheng, Q.: Robust optimization of distance-based tolls in a network considering stochastic day to day dynamics. Transp. Res. Part C: Emerg. Technol. **79**, 58–72 (2017)
12. Liu, Z., Wang, S., Meng, Q.: Optimal joint distance and time toll for cordon-based congestion pricing. Transp. Res. Part B: Methodol. **69**, 81–97 (2014)
13. Liu, Z., Meng, Q., Wang, S.: Speed-based toll design for cordon-based congestion pricing scheme. Transp. Res. Part C: Emerg. Technol. **31**, 83–98 (2013)
14. Yan, R., Wang, S., Fagerholt, K.: A semi-"smart predict then optimize" (semi-SPO) method for efficient ship inspection. Transp. Res. Part B: Methodol. **142**, 100–125 (2020)

Public Transport Passenger Count Forecasting in Pandemic Scenarios Using Regression Tsetlin Machine. Case Study of Agder, Norway

K. Darshana Abeyrathna, Sinziana Rasca, Karin Markvica, and Ole-Christoffer Granmo

Abstract Challenged by the effects of the *COVID-19 pandemic*, public transport is suffering from low ridership and staggering economic losses. One of the factors which triggered such losses was the lack of preparedness among governments and public transport providers. The losses can be minimized if the passenger count can be predicted with a higher accuracy and the public transport provision adapted to the demand in real time. The present paper explores the use of a novel machine learning algorithm, namely *Regression Tsetlin Machine*, in using historical passenger transport data from the current COVID-19 pandemic and pre-pandemic period, combined with a calendar of pandemic-related events (e.g. daily number of new cases and deaths, restrictive measures for pandemic containment), to forecast public transport patronage variations in a pandemic scenario. Results show that the *Regression Tsetlin Machine* has the best accuracy of forecasts when compared to four other models usually employed in the public transport forecasting field. We also observed variations of the prediction accuracy in relation to the period of the pandemic in which the trained models are applied. The underlying reasons for the relative passenger count variations are also examined using the properties of the Tsetlin Machine.

Keywords Public transit · Passenger count forecasting · Pandemic · Regression Tsetlin Machine

K. Darshana Abeyrathna (✉) · O.-C. Granmo
Centre for Artificial Intelligence Research, University of Agder (UiA), N-4898, Grimstad, Norway
e-mail: darshana.abeyrathna@uia.no

O.-C. Granmo
e-mail: ole.granmo@uia.no

S. Rasca
Faculty of Engineering and Science, University of Agder (UiA), N-4898, Grimstad, Norway
e-mail: sinziana.rasca@uia.no

K. Markvica
Center for Mobility Systems, Austrian Institute of Technology (AIT), 1210, Vienna, Austria
e-mail: karin.markvica@ait.ac.at

© The Author(s), under exclusive license to Springer Nature Singapore Pte Ltd. 2021
X. Qu et al. (eds.), *Smart Transportation Systems 2021*, Smart Innovation,
Systems and Technologies 231, https://doi.org/10.1007/978-981-16-2324-0_4

1   Introduction

1.1   The World-wide Situation

The spread of the COVID-19 virus has shed new light on the way urban societies react to pandemic situations. In this new reality, public transport (PT) has an ambivalent character: maintaining the continuation of critical services (hospitals, supermarkets etc.) and being a potential high contamination risk environment due to the typical enclosed and crowded travel conditions. The significant patronage losses triggered by the societal reaction to the pandemic and slow recovery process in ridership point in the direction of a lingering fear of contagion for the population and the need for preparedness for similar future events.

1.2   Case Study of Agder

The region of Agder (approx. 300 000 inhabitants, 80% concentrated in the coastal area) in southern Norway is used as a case study in the present research due to the following reasons: (1) data availability-Agder Kollektivtrafikk AS (AKT), the PT provider in Agder, has granted access to the passenger loads and delays data for their PT lines; (2) replicability-national and international replication potential for spatial and population features; (3) choice of bus line 100-limitation of scope for the model development and initial testing.

1.3   Motivation of Present Research

The likelihood of future pandemics is impossible to predict, but their negative effect on urban mobility is certain, with PT providers struggling to compensate for staggering income losses and market share reduction. Therefore, it is essential to use data from current pandemics in order to prepare the PT network for similar future events by using forecasting models.

A promising way to identify patterns in lockdown effects on the PT network is to employ machine learning (ML) algorithms in analyzing big data sets for forecasting purposes. In this study, we use the Regression Tsetlin Machine (RTM) [1], a variant of the binary Tsetlin Machine (TM) [2], to predict the variation in PT ridership in pandemic scenarios for viruses that are transmitted in a similar way to COVID-19. The main advantage of using TMs is the competitive accuracy of predictions, despite the fact that they are made of interpretable rule-based classifiers, memory footprint, and inference speed.

The RTM has been previously used to predict e.g. the dengue incidences in the Philippines [3]. Abeyrathna et al. in [4–6], discussed the interpretability of the binary TM on distinct applications. In our study, this algorithm is applied for the first time to the transport domain, where we predict the PT ridership variation during the pandemic period providing evidence on the interpretability of the TM based approach through generating rules, which can be used for travel behavior predictions in future pandemic scenarios. The results show immediate application potential for PT providers who could use the model for adapting their PT provision to the actual market demand, especially in the case of regional and national lockdowns.

2 State of the Art

2.1 Role of Public Transport in a Pandemic

Browne et al. [7] studied the relationship between the spread of respiratory viruses and PT, revealing that the duration of travel and seating proximity influences the risk of infection for ground transport. In terms of patronage, a study of the 2002/03 SARS pandemic [8] showed that reported new cases caused immediate losses in the underground ridership, even with no lockdown in place ("fresh fear") [9]. When studying European mobility data, Santamaria et al. [10] found that "confinement measures explain up to 90% of the mobility patterns".

2.2 Machine Learning in Public Transport Research

Currently, applications of ML in PT research are popular in the domains of:

- Travel mode choice modeling-analyzing user data to accurately predict mode choice. Previous research indicates ML models, i.e. random forest or different artificial neural nets, as the best performers in the field [11–13].
- Travel demand modelling and forecasting-models such as Autoregressive Integrated Moving Average, dynamic Partial Adjustment Model have been used by Chi-Hong et al. [14] to predict PT demand. Mozolin et al. used neural networks (NN) for trip distribution forecasting [15]. Koushik et al. discuss that the results of applying NN for travel demand models are not in favour of NN due to the "black-box" effect [16].
- Forecasting passenger flows-Toque et al. used gated recurrent unit and recurrent NN for short term prediction of passenger flows, comparing the results with the ones of Random Forest and long-term forecasting models [17]. Wei and Chen employed the empirical mode decomposition and NN [18] for the same purpose. Toque et al. extended their previous study to long-term forecasting as well [19].

2.3 Machine Learning in the COVID-19 Pandemic

In a 2020 review, Lalmuanawma et al. conclude that "the ongoing development in AI and ML has significantly improved treatment, medication, screening, prediction, forecasting, contact tracing, and drug/vaccine development process for the Covid-19 pandemic and reduced the human intervention in medical practice. However, most of the models are not deployed enough to show their real-world operation" [20]. Another example is predicting the number of new cases for COVID-19 [21]. A recent interpretable ML algorithm, the Regression Tsetlin Machine based mobile application, has been developed for the same purpose [22].

3 Methodology

In this section, we will briefly introduce the theory of the ML algorithm employed in this study, the data set used, the data pre-processing approach, and the model validation technique. Figure 1 presents the diagram of the planned work-flow.

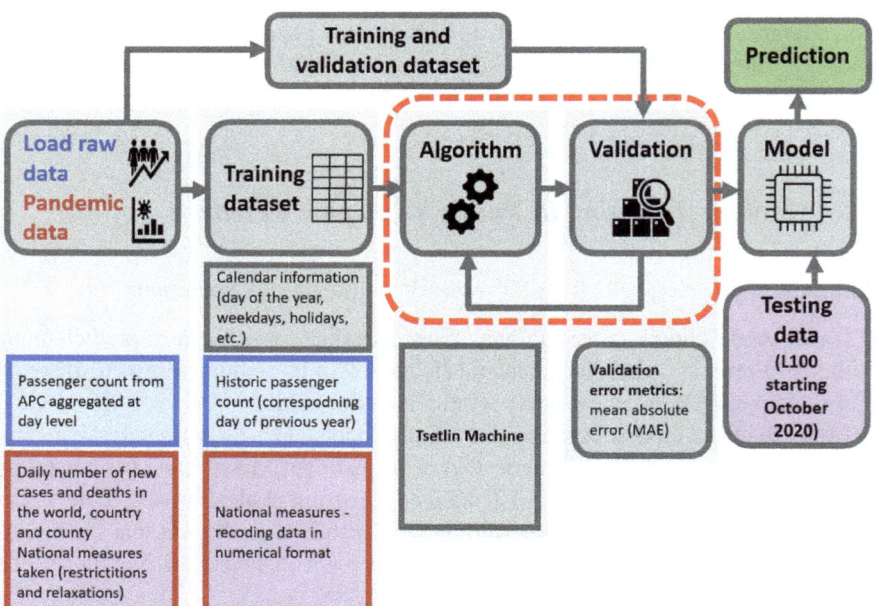

Fig. 1 Method concept

3.1 The Regression Tsetlin Machine Algorithm

The RTM is a variation of TMs and a novel approach to interpretable non-linear regression [1]. The RTM takes o propositional input features, i.e., $\mathbf{X} = [x_1, x_2, x_3, \ldots, x_o,]$ and sends them along with their negations (collectively called literals), i.e., $\mathbf{X}' = [x_1, x_2, x_3, \ldots, x_o, \neg x_1, \neg x_2, \neg x_3, \ldots, \neg x_o]$ to each of the clauses, c_j, $j = 1, 2, 3, \ldots, m$. Each clause comprises of a team of Tsetlin Automata (TAs) which decides the composition of the clause. Individual TAs attached to each literal decide to include or exclude their corresponding literals in the clause. The clause then computes the conjunction of only the included literals in the clause. By carefully guiding these TAs to make correct decisions as a team, individual clauses recognize the sub-patterns in data.

Once a sub-pattern is recognized by a clause, the clause outputs 1. The resulting sum of the individual clause outputs is then mapped into a continuous value between 0 and $\hat{y}_{\max}$, where $\hat{y}_{\max}$ is the maximum training output.

During the training phase, the predicted daily passenger count, y, is compared against the actual passenger count output, $\hat{y}$. Depending on whether y (predicted value) is higher or lower than $\hat{y}$, the individual clauses are systematically guided to reduce the prediction error.

3.2 Data Set

The dataset was compiled from various sources:

- passenger count data for bus Line 100, made available by AKT. The data was collected internally by AKT using Automated Passenger Counting (APC). In our dataset we used aggregated daily values for Line 100.
- daily number of newly registered COVID-19 cases and deaths in the world (data sourced from www.ourworldindata.org), Norway and Agder (data sourced from the National Institute of Public Health in Norway).
- school holiday schedule in Agder, available at www.publicholidays.no.
- restrictive and relaxation measures imposed nationally in Norway in relation to the COVID-19 pandemic containment strategy. The data was collected from the official websites of various Norwegian ministries.

3.3 Features, Predictions, and Data Preprocessing

What we expect in this study is to train the RTM algorithm using historical PT ridership data from bus line 100 and the calendar of pandemic related events to predict the PT daily ridership variation in future pandemic situations. The model requires the input of different features to be able to estimate the future passenger count for a similar situation.

In our case, the passenger count, PC of day d for the bus line 100, $PC(d)$ is predicted using the previous day passenger count, $PC(d-1)$ previous year-same-day passenger count of bus line 100, $PC(d-365)$, previous day passenger count of the Agder province, $PC_A(d-1)$, previous year-same-day passenger count of the Agder province, $PC_A(d-365)$, number of new corona cases of day $d-1$ worldwide, $NCC_w(d-1)$ number of corona cases of day $d-1$ in Agder, $CC_A(d-1)$, number of corona cases of day $d-1$ in Norway, $CC_N(d-1)$, number of corona cases of day $d-1$ worldwide, $CC_w(d-1)$, number of corona deaths of day $d-1$ in Agder, $CD_A(d-1)$, number of corona deaths of day $d-1$ in Norway, $CD_N(d-1)$, number of corona deaths of day $d-1$ worldwide, $CD_w(d-1)$, holiday information related to day d, and different pandemic related measures related to day d. In the first phase of preprocessing, the non-numerical features (e.g. pandemic measures) are encoded to numerical values without losing the true meaning of the feature. In the second phase, the complete set of numerical features are binarized as the RTM accepts only binary form features. For feature binarization, we use the thresholding approach proposed in [3].

3.4 Training and Validation

The RTM model is trained on different training samples by varying the day d between 1 January 2020 and day t. In this study we focus on medium-term prediction of passenger count: up to two weeks. Hence, once the model has been trained on the data from 1 January 2020 to day t, this model is used to predict the passenger count for the days from $t+1$ to $t+14$.

However, for real-life situations, predictions of passenger count for the above days have to be made based on the predictions of $PC(d-1)$, PC_A, CC_A, CC_N, NCC_w, CC_w, CD_A, CD_N, CD_w, and pandemic related measures. For instance, to predict the passenger count of day $t+7$, $PC(t+7)$ the model will require the feature values of some of the features of day $t+6$, which are not available by day t. Hence, we use simple moving average approach to predict those input features and then send them to the model to make predictions for the future.

The pandemic related measures also have to be estimated to make prediction for the next two weeks. Since numerical analysis can not be used to estimate the future pandemic related measures, we consider three cases where we assume that, for the next 14 days (1) no pandemic measure is taken, (2) one restrictive pandemic measure is taken, and (3) two restrictive pandemic measures are taken. For cases (2) and (3), the day of the measure is randomly selected. Considering the above conditions, the RTM is used to make two predictions: Case 1 and Case 2. For Case 1, the RTM is trained on data from 1/1/2020 to 9/15/2020 and predictions made for 9/16/2020 to

9/29/2020. For Case 2, the RTM is trained on the data from 1/1/2020 to 10/8/2020 and predictions made for 10/9/2020 to 10/22/2020. We will further present the accuracy values of the predictions.

4 Results and Discussion

The performance of the RTM in relation to the above experiment is measured in terms of the mean absolute error (MAE) between actual and predicted passenger counts. We also contrast the performance of the RTM with a classic statistical model, Moving Average (MA), and three other widely used machine learning models: Random Forest (RF), Regression Trees (RT), and Support Vector Machine (SVM). Their performance on Cases 1 and 2 is summarized in Tables 1 and 2.

Regardless of the number of measures used in testing, the RTM obtains the lowest MAE in both cases. On average, RT exhibits second best performance, closely followed by RF, while SVM and MA struggle to make competitive predictions. The data shows that the error is higher for the Case 1 testing (Figs. 2 and 3). Figure 2 also shows that all machine learning models barely recognize the ridership reduction in the weekends.

A possible reason for the above observation could be the lesser number of training samples the Case 1 has compared to Case 2, as Case 1 consists of 259 training sample while Case 2 contains 282. In our case, the missing samples in Case 1 are highly important since this seems to be the period where the impact of the second pandemic wave starts to be visible on the ridership of Line 100.

Table 1 MAE between actual and predicted passenger counts (testing for Case 1)

		Method				
		RTM	RF	RT	SVM	MA
No. of measures	0	338.41	376.82	370.76	490.78	495.88
	1	337.81	375.50	370.62	490.79	495.88
	2	337.13	375.41	354.74	490.80	495.88

Table 2 MAE between actual and predicted passenger counts (testing for Case 2)

		Method				
		RTM	RF	RT	SVM	MA
No. of measures	0	192.48	217.66	193.75	401.13	445.64
	1	187.16	237.29	193.08	401.10	445.64
	2	186.89	234.26	192.79	401.09	445.64

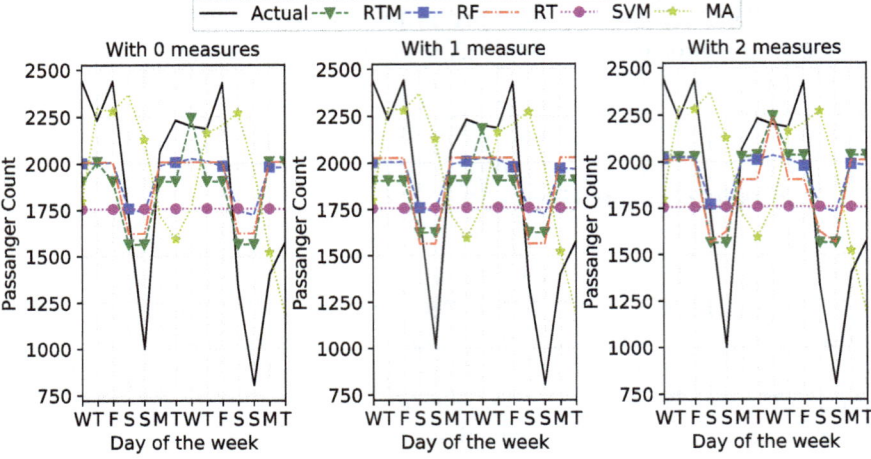

Fig. 2 Actual versus predicted passenger count for the testing in Case 1

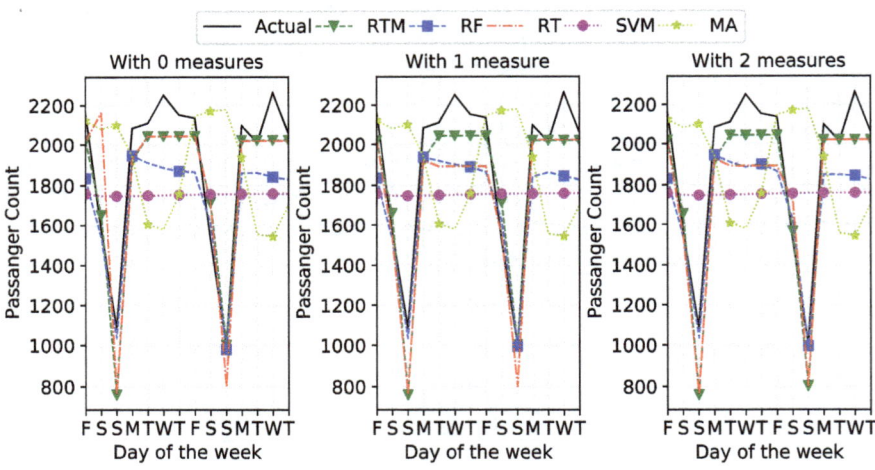

Fig. 3 Actual versus predicted passenger count for the testing in Case 2

Since the drastic pandemic control measures in the first wave had been different in relation to the number of new daily cases in Norway, the available 259 training samples did not provide enough similar samples for the ML training in this situation. In the lockdown period of the first wave (March-April), the passenger count difference between weekdays and weekends is much smaller than for the two cases analyzed. Therefore, in Case 1 the ML models predict similar patterns to the lockdown period, with smaller passenger count differences between weekdays and weekends. For Case 2, the models learn to recognize the differences with better accuracy due to the increased number of training samples. Another possible reason could be the predicted input data for the testing period. For Case 1 the predicted values have been mainly derived from data previous to the second pandemic wave. For Case 2 the data used for predictions is already overlapping the second pandemic wave, hence containing more realistic values as testing inputs than in Case 1.

A crucial result of the study is generating applicable rules for ridership variation predictions by interpreting the logical outputs given by the TM results. Hence, we changed the output in our dataset from regression to categorical and then applied the binary TM on it. The output is categorized as "positive trend" with a passenger count higher than the average of last 14 days, and "negative trend" otherwise. This way the clauses in the TM identify the reasons for positive and negative trends.

Using a similar approach to the ones in [4–6], we identify the reasons for a positive trend using the complete set of features as,

IF $CD_A(d-1) \leq 7390$ **AND** $CD_N(d-1) \leq 9$ **AND** $PC(d-365) > 2220$ **AND** $PC(d-1) > 955$ **AND** $PC_A(d-1) > 6908$ **AND** $PC_A(d-365) > 39401$ **AND** day d is **NOT** a holiday **AND** Days to enforce a measure ≤ 5 **AND** day d is a week day **THEN** a positive trend.

Similarly, the reasons for a negative trend is identified as,

IF $PC(d-365) \leq 1997$ **AND** $PC_A(d-365) \leq 33703$ **AND** No pandemic control relaxation measures taken **AND** No public transport restrictive measures taken **AND** day d is a weekend day **THEN** a negative trend.

Using the above rule, the correct trend can be predicted with an accuracy of over 86%.

The next step is to remove the number of cases and deaths worldwide, and the historical ridership data from the features set and re-run the model. The resulting rule generated from the remaining features still predicts the trend with an accuracy close to 85%. Thus, the reasons for a positive trend using the filtered set of features are identified as,

IF $CC_N(d-1) \leq 100$ **AND** the number of days after a positive announcement ≤ 38 **AND** day d is a week day **THEN** a positive trend.

Similarly, the reasons for a negative trend is identified as,

IF day d is a weekend day **THEN** a negative trend.

5 Conclusions and Future Research

Our research concentrates on the application of RTM on forecasting ridership variation in PT in the specific conditions of a pandemic where the virus spreads similarly to COVID-19. The results in Tables 1 and 2 show that RTM obtains the lowest mean absolute error for forecasting the variation in PT ridership in comparison to all other

ML models tested (RF, RT, SVM, MA). Evidence on the interpretability of the TM is also given, allowing for the formulation of forecasting rules.

Therefore, the method presents good potential for supporting PT providers and decision makers in their response to pandemic scenarios that affect PT ridership. Enabling continuous learning by implementing current data on patronage, timetable alterations and local restrictions, allows for high accuracy ridership variation forecasting which can foster a real-time adaptation of PT provision. This approach could help minimise financial and patronage losses for PT providers and public authorities in the events of severe patronage reductions, such as lockdowns.

When discussing the accuracy of prediction, we observe variations in relation to the period of the pandemic for which simulations are being run. For the second wave, the accuracy of the prediction depends on how far within the second wave the simulation is being run. It may be necessary to correct the data in future pandemic scenarios (i.e. correct input data set to train model on similar number of daily cases) to ensure accuracy.

From the generated rules, we observe that only marking a pandemic control measure as positive or negative does not correctly estimate the impact strength of the measure itself. Therefore, further research is necessary on this topic.

References

1. Abeyrathna, K.D., Granmo, O.-C., Zhang, X., Jiao, L., Goodwin, M.: The regression tsetlin machine: a novel approach to interpretable nonlinear regression. Philosoph. Trans. Royal Soc. A **378**(2164), 20190165 (2020)
2. Granmo, O.-C.: The tsetlin machine-a game theoretic bandit driven approach to optimal pattern recognition with propositional logic.' *arXiv preprint* arXiv:1804.01508 (2018)
3. Abeyrathna, K.D., Granmo, O.-C., Zhang, X., Goodwin, M.: A scheme for continuous input to the tsetlin machine with applications to forecasting disease outbreaks. In: International Conference on Industrial, Engineering and Other Applications of Applied Intelligent Systems. Springer, pp. 564–578 (2019)
4. Abeyrathna, K.D., Granmo, O.-C., Goodwin, M.: Extending the tsetlin machine with integer-weighted clauses for increased interpretability. *arXiv preprint* arXiv:2005.05131 (2020)
5. Abeyrathna, K.D., Pussewalage, H.S.G., Ranasinghe, S.N., Oleshchuk, V.A., Granmo, O.-C.: Intrusion detection with interpretable rules generated using the tsetlin machine. In: 2020 IEEE Symposium Series on Computational Intelligence (SSCI). IEEE (2020)
6. Abeyrathna, K.D., Granmo, O.-C., Goodwin, M.: On obtaining classification confidence, ranked predictions and auc with tsetlin machines. In:2020 IEEE Symposium Series on Computational Intelligence (SSCI). IEEE (2020)
7. Browne, A., St-Onge Ahmad, S., Beck, C.R., Nguyen-Van-Tam, J.S.: The roles of transportation and transportation hubs in the propagation of influenza and coronaviruses: a systematic review. J. Travel Med. **23**(1), tav002 (2016)
8. Chen, K.T., Twu, S.J., Chang, H.L., Wu, Y.C., Chen, C.T., Lin, T.H., Olsen, S.J., Dowell, S.F., Su, I.J., Team: SARS in Taiwan: an overview and lessons learned. Int. J. Infect. Dis. **9**(2), 77–85 (2005)
9. Wang, K.-Y.: How change of public transportation usage reveals fear of the SARS virus in a city. PloS one **9**(3) (2014)

10. Santamaria, C., Sermi, F., Spyratos, S., Iacus, S.M., Annunziato, A., Tarchi, D., Vespe, M.: Measuring the impact of covid-19 confinement measures on human mobility using mobile positioning data. a european regional analysis. Safety Sci. **132** (2020)
11. Zhao, X., Yan, X., Yu, A., Van Hentenryck, P.: Prediction and behavioral analysis of travel mode choice: a comparison of machine learning and logit models. Travel Behav. Soc. **20**, 22–35 (2020)
12. Cheng, L., Chen, X., De Vos, J., Lai, X., Witlox, F.: Applying a random forest method approach to model travel mode choice behavior. Travel behav. Soc. **14**, 1–10 (2019)
13. Hagenauer, J., Helbich, M.: A comparative study of machine learning classifiers for modeling travel mode choice. Expert Syst. Appl. **78**, 273–282 (2017)
14. Tsai, C.-H.P., Mulley, C., Clifton, G., et al.: Forecasting public transport demand for the Sydney greater metropolitan area: a comparison of univariate and multivariate methods. Road Transp. Res. J. Australian and New Zealand Res. Pract. **23**(1), 51 (2014)
15. Mozolin, M., Thill, J.-C., Usery, E.L.: Trip distribution forecasting with multilayer perceptron neural networks: a critical evaluation. Transport. Res. Part B Methodol. **34**(1), 53–73 (2000)
16. Koushik, A.N., Manoj, M., Nezamuddin, N.: Machine learning applications in activity-travel behaviour research: a review. Transp. Rev. **40**(3), 288–311 (2020)
17. Toqué, F., Côme, E., Oukhellou, L., Trépanier, M.: Short-term multi-step ahead forecasting of railway passenger flows during special events with machine learning methods (2018)
18. Wei, Y., Chen, M.-C.: Forecasting the short-term metro passenger flow with empirical mode decomposition and neural networks. Transport. Res. Part C Emerging Technol. **21**(1), 148–162 (2012)
19. Toqué, F., Khouadjia, M., Come, E., Trepanier, M., Oukhellou, L.: Short & long term forecasting of multimodal transport passenger flows with machine learning methods. In: 2017 IEEE 20th International Conference on Intelligent Transportation Systems (ITSC), pp. 560–566. IEEE (2017)
20. Lalmuanawma, S., Hussain, J., Chhakchhuak, L.: Applications of machine learning and artificial intelligence for covid-19 (sars-cov-2) pandemic: a review. Chaos, Solitons & Fractals, p. 110059 (2020)
21. Tuli, S.,Tuli, S., Tuli, R., Gill, S.S.: Predicting the growth and trend of covid-19 pandemic using machine learning and cloud computing. Internet of Things, p. 100222 (2020)
22. Abeyrathna, K.D.: The regression tsetlin machine based ai enabled mobile app for forecasting the number of corona patients for the next day in different countries. GitHub repository (2019)

Travel Time Reliability Analysis of Arterial Road Based on Burr Distribution

Yixiao Lu and Fujian Wang

Abstract Travel time distribution has been widely used to characterize the arterial road traffic conditions and help to analyze travel time reliability. Based on automatic number plate recognition (ANPR) data, this paper studies the travel time distribution and reliability of urban arterial roads and analyzes the relationship between the number of intersections and travel time characteristics. Firstly, the ANPR data is fitted by Burr distribution, and conducted the goodness-of-fit tests. Then, the coefficient of variation, buffer index and planning time index are used as reliability evaluation indexes to evaluate the travel time reliability. The results indicate that Burr distribution has a high acceptance rate. With the increase of the number of intersections, the route travel time distribution tends to a stable unimodal distribution, and the goodness of fit of Burr distribution increases, and the values of the three reliability evaluation indexes decrease, indicating that the travel time fluctuation decreases and the reliability increases.

Keywords Burr distribution · Travel time reliability · Signalized intersection

1 Introduction

Travel time distribution and its reliability have been widely used to evaluate the traffic operation condition and transportation network performance. From the perspective of managers, the analysis of travel time is an important basis to support traffic planning, design and management policy-making. As for travelers, acquiring information about travel time could be useful when making travel plans, choosing departure time and planning travel routes.

On the study of travel time distribution, the initial view was that normal distribution was appropriate. Later, Wardrop [1] proposed that travel time obeyed skew distribution. Herman and Lam [2] analyzed the travel time data of urban arterial roads in Detroit, found that there was obvious deviation in travel time, and proposed

Y. Lu (✉) · F. Wang
College of Civil Engineering and Architecture, Zhejiang University, Hangzhou 310058, China
e-mail: luyixiao0501@zju.edu.cn

© The Author(s), under exclusive license to Springer Nature Singapore Pte Ltd. 2021   39
X. Qu et al. (eds.), *Smart Transportation Systems 2021*, Smart Innovation,
Systems and Technologies 231, https://doi.org/10.1007/978-981-16-2324-0_5

Gamma distribution and lognormal distribution to express the variability of travel time. Richardson and Taylor [3] collected and analyzed the travel time data of main roads in Melbourne, and observed that the travel time variability can be expressed by lognormal distribution. Taylor et al. [4] analyzed the travel time data of two main roads in Adelaide, and found that Burr distribution can better represent the travel time variability than lognormal distribution, and Burr distribution also provides a computational advantage for the use of percentage-based reliability evaluation index. Susilawati [5] subsequently showed that Burr distribution can fit most of the observed travel time variability data. Zhen [6] analyzed the travel time distribution in different periods, different weeks, different locations and different weather conditions, and the results showed that Burr distribution can better describe the travel time.

The traffic flow of urban road is more complex, affected by signalized intersections, entrances and exits, crossing pedestrians and other factors, showing multi peak characteristics. Compared with normal distribution and lognormal distribution, some studies show [7–9] that the bimodal probability distribution model can better reflect the travel time and speed characteristics of urban interrupted flow. Taylor and susilawati [10] verified that the arterial travel time is correspond to bimodal distribution. However, they further found that the bimodal probability of travel time distribution decreases when the short links are merged into a long single route under the same signal control system, and the bimodal phenomenon will be broken up. Luo [11] proposed that the multi peak state in short periods and the unimodal state in long periods may coexist. Similarly, Liu [12] found through data fitting that with the increase of distance and the number of intersections, the bimodal distribution phenomena weakened, and the impact of red light delay on travel time distribution decreased.

Travel time reliability evaluation index is a measurement of reliability analysis, which has important practical significance for managers and travelers. Standard deviation and coefficient of variation are commonly used to describe travel time reliability [13]. At the same time, some studies [14] pointed out that percentile values can be used as reliability measures, such as 90th or 95th percentile values, and indicators calculated based on percentile values, such as buffer index, skewness and width of travel time. Most of the mathematical expressions of the above indexes are based on percentiles can also be used as travel time reliability measures.

Based on the ANRP data in Xiaoshan District of Hangzhou, this paper studies the distribution characteristics of arterial road travel time by using Burr distribution. Considering that the percentile value of Burr distribution is easy to calculate, the coefficient of variation, buffer index and planned time index are selected as the reliability index for reliability analysis. Furthermore, the relationship between the number of intersections in the road and the travel time characteristics of arterial road is analyzed.

2 Data Description

2.1 Data Source

Electronic police cameras are installed at signalized intersections in Xiaoshan District of Hangzhou, which can collect information such as vehicle license plates and the time of passing through the stop line. The correct rate of recognizing the complete license plate is more than 95%, and the detection time can be accurate to seconds [15], therefore, the information can be the data basis of the research. Through license plate number matching method, the travel time of vehicles passing through two continuous intersections can be calculated, and the travel time of vehicles on the target road can be obtained.

The electronic police camera is generally set at about 20 m upstream of the stop line at the entrance of signalized intersection to capture the video information of vehicles passing through the stop line. Therefore, in this study, the link travel time refers to the running time of vehicles from the stop line of upstream intersection to the stop line of downstream intersection.

This study selects a signal coordinated control route in Xiaoshan District: from the intersection of Shixin road and Jianshe fourth road to the intersection of Shixin road and Bengjing Avenue. The route includes 8 signalized intersections with a total length of 4.4 km (Fig. 1).

Fig. 1 Schematic diagram of intersection location and number

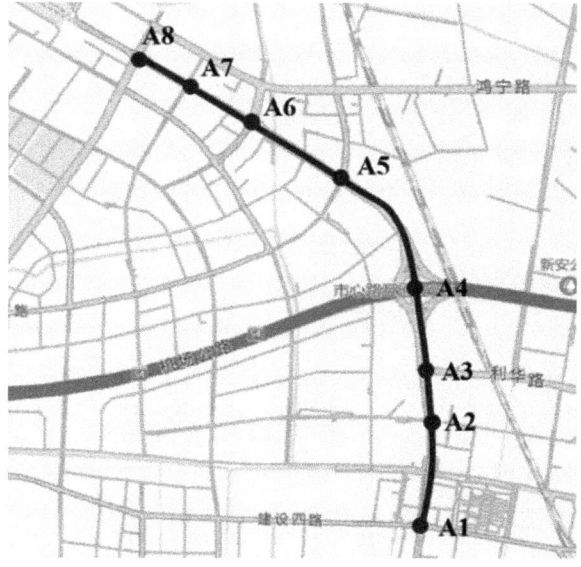

Table 1 Link statistical information

Link number	Link length (m)	Sample size	Mean (s)	Median (s)	Standard deviation
A1-A2	758	12,857	93.42	73	40.51
A2-A3	352	17,776	50.15	33	74.58
A3-A4	581	7911	83.28	79	42.07
A4-A5	994	5736	112.26	105	67.69
A5-A6	741	10,466	81.76	66	45.06
A6-A7	521	15,549	63.31	47	41.43
A7-A8	396	14,218	67.33	50	48.95

2.2 Basic Statistics of Link Travel Time

Based on the ANPR data of South to North traffic flow at the morning peak in March 2019, calculate the link length, the sample size, the travel time mean, median and standard deviation value of each link, as shown in Table 1.

3 Travel Time Distribution Fitting

3.1 Burr Distribution

Burr distribution is widely used in medicine, engineering, finance and insurance. Susilawati [10] applied Burr distribution to travel time distribution fitting, and found that its fitting effect was better than Gamma distribution and Weibull distribution. The expressions of probability density function and cumulative distribution function of 3-parameter Burr (type XII) distribution are as follows [16]:

$$f(x|\alpha, c, k) = ck(x/\alpha)^{c-1}(1 + (x/\alpha)^c)^{-(k+1)} \tag{1}$$

$$F(x, c, k) = 1 - (1 + (x/\alpha)^c)^{-k} \tag{2}$$

where, α is the scale parameter, k is the shape parameter, which is used to describe the shape change of the curve, and c is the position parameter, which is used to describe the position change of the curve, especially the position of the peak value of the curve.

The mean of Burr distribution is calculated as follows:

$$mean = E(X) = \alpha k \frac{\Gamma\left(k - \frac{1}{c}\right)\Gamma\left(1 + \frac{1}{c}\right)}{\Gamma(k + 1)} \tag{3}$$

where, Γ is Gamma function, the functional form is $(t) = \int_0^{+\infty} x^{t-1} e^{-x} dx$.
The standard deviation calculation formula is as follows:

$$SD = \sqrt{\alpha^2 k \frac{\Gamma\left(k - \frac{2}{c}\right)\Gamma\left(1 + \frac{2}{c}\right)}{\Gamma(k+1)} - \left(\alpha k \frac{\Gamma\left(k - \frac{1}{c}\right)\Gamma\left(1 + \frac{1}{c}\right)}{\Gamma(k+1)}\right)^2} \tag{4}$$

The percentile can be obtained from the cumulative distribution function:

$$x_p = \alpha \sqrt[c]{(1 - p)^{-\frac{1}{k}} - 1} \tag{5}$$

3.2 Distribution Fitting Results and Parameters Estimation

There are eight signalized intersections in the selected arterial road, in order to study the influence of the number of intersections on the travel time characteristics, taking intersection A1, A2, A3, A4, A5 and A6 as the origin respectively, the route travel time data with different number of intersections are selected, so there are 26 groups of route travel time data with different number of intersections.

Burr distribution is used to fit the link and route travel time data, and goodness of fit test is conducted. The decision value H = 0 means to accept the hypothesis of passing the test; H = 1 means to reject the hypothesis of passing the test. Then the maximum likelihood estimation method is used to estimate the parameters of Burr distribution. The results are shown in Table 2.

According to the fitting results of Burr distribution, at the 5% significance level, 18 groups of data pass the goodness of fit test, and the acceptance rate is 69.23%. Burr distribution is rejected by most of the travel time data of the route with less number of intersections or short length, and all single link travel time distributions rejected Burr distribution. This is because when the number of intersections is less, the delay caused by intersection signal control has a great impact on the travel time. At this time, the travel time distribution presents a bimodal phenomenon or even multi peak phenomenon, and multi-mode distribution can be tested better than unimodal distribution. With the increase of the number of intersections, the travel time tends to a stable unimodal distribution, and the goodness of fit of Burr distribution increases, and Burr distribution can better describe the characteristics of high peak and right deviation of travel time distribution.

Figure 2 shows the fitting diagram of travel time distribution of route with different number of intersections. It can be found that, with the increase of the number of intersections and the length of route, the travel time data has obvious characteristics of long right tail, while Burr distribution can better describe the characteristics of high peak value and right deviation of data.

Table 2 Parameter estimation and goodness of fit test results of Burr distribution

Route origin	Number of intersections	Route length	Burr distribution		Burr distribution parameter estimation		
			H	P	α	c	k
A1	2	758	1	0.0	67.915	7.1787	0.4377
	3	1110	1	0.0	122.26	9.8972	0.2541
	4	1691	0	0.0723	254.535	5.76273	0.8295
	5	2685	0	0.2574	324.88	6.0446	0.7311
	6	3426	0	0.6568	363.185	8.4868	0.5294
	7	3947	0	0.6185	351.434	−9.4581	2.6283
A2	2	352	1	0.0	43.715	13.4785	0.1360
	3	933	0	0.1207	172.146	4.2058	0.8437
	4	1927	0	0.3079	206.181	5.6217	0.5635
	5	2668	0	0.9324	271.136	8.2778	0.4252
	6	3189	0	0.1248	349.458	7.1275	0.6146
	7	3585	0	0.4203	396.897	7.5122	0.4844
A3	2	581	1	0.0148	120.357	2.9643	2.3292
	3	1575	0	0.3399	126.459	5.5024	0.6108
	4	2316	0	0.5243	189.223	8.6877	0.4203
	5	2837	0	0.6234	266.853	7.9622	0.5074
	6	3233	0	0.5418	354.783	7.1240	0.6375
A4	2	994	1	0.0	55.7272	4.3358	0.7713
	3	1735	0	0.0907	106.842	11.0711	0.3012
	4	2256	0	0.4550	154.255	19.0948	0.1530
	5	2652	0	0.5108	211.728	26.0105	0.0932
A5	2	741	1	0.0	39.4249	11.6059	0.1712
	3	1262	1	0.0	166.115	4.2120	1.4123
	4	1658	0	0.1740	220.461	4.9294	1.2217
A6	2	521	1	0.0	38.8604	12.3952	0.1438
	3	917	0	0.0906	237.751	2.9992	2.1857

4 Travel Time Reliability

4.1 Travel Time Reliability Index

The evaluation indexes of travel time reliability include skewness (λ^{skew}), coefficient of variation (CV), buffer index (BI), travel time index (TTI), planning time index (PI), etc. Because Burr distribution is flexible in mathematics, it is easy to calculate its percentile value, so this paper selects the coefficient of variation, buffer index and planning time index as the reliability evaluation index to calculate, the calculation

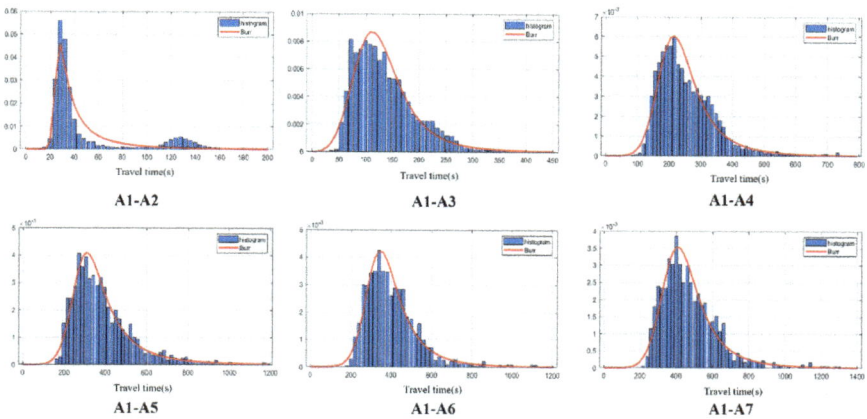

Fig. 2 Burr distribution fitting diagram for different number of intersections

formula is as follows:

$$CV = \frac{\sigma}{\mu} = \frac{\alpha k \frac{\Gamma\left(k-\frac{1}{c}\right)\Gamma\left(1+\frac{1}{c}\right)}{\Gamma(k+1)}}{\sqrt{\alpha^2 k \frac{\Gamma\left(k-\frac{2}{c}\right)\Gamma\left(1+\frac{2}{c}\right)}{\Gamma(k+1)} - \left(\alpha k \frac{\Gamma\left(k-\frac{1}{c}\right)\Gamma\left(1+\frac{1}{c}\right)}{\Gamma(k+1)}\right)^2}} \tag{6}$$

$$BI = \frac{t_{95\%} - t_{50\%}}{t_{50\%}} = \alpha \frac{\sqrt[c]{20^{1/k} - 1} - \sqrt[c]{2^{1/k} - 1}}{\sqrt[c]{2^{1/k} - 1}} \tag{7}$$

$$PI = \frac{t_{95\%}}{t_{15\%}} = \frac{\sqrt[c]{20^{1/k} - 1}}{\sqrt[c]{\left(\frac{20}{17}\right)^{1/k} - 1}} \tag{8}$$

According to the Burr distribution parameters obtained by the maximum likelihood method and Eqs. (6)–(8), the reliability indexes are calculated. The results are shown in Table 3.

According to the calculation results of reliability indexes, in the six groups of data, the reliability indexes show a similar trend. With the increase of the number of intersections and the length of route, the coefficient of variation, buffer index and planning time index show a downward trend, indicating that the reliability increases. The reason is analyzed: when vehicles pass through a single link, they are vulnerable to the influence of intersection signal control, especially in the morning peak period, red light waiting time may account for a large proportion of travel time. The BI values of single link travel time are greater than 1 or even 2, and the PI value are greater than 3, which indicates that the travel time is extremely unreliable, so the travel time reliability of a single link is least reliable. When the road length is longer than 3 km, BI will be less than 1 or 0.5, PI will be less than 2 or 2.5, indicating that the travel time is slightly unreliable, compared with single link, the reliability degree increases.

Table 3 Calculation results of reliability index

Route origin	Number of intersections	Length of route	CV	BI	PI
A1	2	758	0.5009	1.1482	2.8997
	3	1110	0.7490	1.5153	3.3278
	4	1691	0.4520	0.7781	2.4291
	5	2685	0.4413	0.8209	2.4723
	6	3426	0.3455	0.7318	2.1967
	7	3947	0.3028	0.2992	1.6003
A2	2	352	0.7051	1.5134	3.8179
	3	933	0.5980	1.1808	3.3389
	4	1927	0.5535	1.1978	3.1258
	5	2668	0.4114	0.9747	2.5686
	6	3189	0.3940	0.7850	2.3407
	7	3585	0.4190	0.9520	2.5738
A3	2	581	0.6048	0.9782	3.3572
	3	1575	0.5299	1.1260	3.0219
	4	2316	0.3872	0.9252	2.4762
	5	2837	0.3773	0.8341	2.3714
	6	3233	0.3610	0.7567	2.0977
A4	2	994	0.6085	1.2351	3.4056
	3	1735	0.3979	1.0138	2.5310
	4	2256	0.5010	1.2005	2.6966
	5	2652	0.4701	1.5853	3.2417
A5	2	741	0.7355	1.1911	3.3409
	3	1262	0.5457	0.7889	2.6454
	4	1658	0.5006	0.7055	2.3982
A6	2	521	0.8094	1.6412	4.0581
	3	917	0.6115	0.9896	3.3647

With the increase of route length, the probability of vehicles affected by intersection signal control decreases, and the proportion of red light waiting time in travel time decreases. Therefore, the value of coefficient of variation, buffer index and the planning time index show a downward trend, at the same time, travel time reliability increases. It can be concluded that with the increase of road length and the number of intersections, the volatility of travel time decreases, the reliability is increased.

5 Conclusions

In order to analyze the travel time characteristics of urban arterial road, this paper uses the travel time data with different number of intersections based ANPR data, and analyzes the travel time distribution and reliability characteristics. Firstly, the goodness of fit of travel time data is tested based on Burr distribution, and then analyzed the travel time reliability by coefficient of variation, buffer index and planning time index. The conclusions of this paper can be concluded as follows.

Burr distribution is a useful model to represent the route travel time reliability, showing a high acceptance rate, and can well describe the characteristics of right deviation and high peak of travel time. Further considering the number of intersections and the length of routes, it is found that the travel time of single link and short routes often presents a bimodal phenomenon, which may be caused by the delay of intersection signal control. However, with the increase of the number and length of routes, the travel time distribution tends to be a stable unimodal distribution, showing the characteristics of long right tail. And the more the number of intersections and the longer the route length, the higher the goodness of fit of Burr distribution. It shows that with the increase of route length, the influence of intersection signal control is weakened and the travel time reliability increases.

Coefficient of variation, buffer index and coefficient of variation are used as reliability evaluation indexes to evaluate the travel time reliability characteristics of different road lengths. The results show that the travel time of a single link are extremely unreliable. With the increase of the number of intersections and the length of road, the values of the three indicators show a downward trend, the travel time reliability increase.

References

1. Wardrop, J.G.: Some theoretical aspects of road traffic research. Proc. Inst. Civ. Eng. **1**(2), 325–379 (1952)
2. Herman, R., Lam, T.: Trip time characteristics of journeys to and from work. Transp Traffic Theory **6**, 57–86 (1974)
3. Richardson, A.J., Taylor, M.A.P.: Travel time variability on commuter journeys. High Speed Ground Transp J **12**(1), 77–99 (1978)
4. Taylor, M.A.P., Susilawati: Modelling travel time reliability with the burr distribution. Procedia Soc. Behav. Sci. **54**(2290), 75–83 (2012)
5. Susilawati, S., Taylor, M.A., Somenahalli, S.V.: Distributions of travel time variability on urban roads. J. Adv. Transp. **47**(8), 720–736 (2013)
6. Zhen, C., Wei, D.: Analyzing travel time distribution based on different travel time reliability patterns using probe vehicle data. Int. J. Transp. Sci. Technol. (2020)
7. Yang, F., Yun, M., Yang, X.: Travel time distribution under interrupted flow and application to travel time reliability. Transp. Res. Rec. J. Transp. Res. Board **2466**(2466), 114–124 (2014)
8. Li, R., Qian, X., Wu, H.: Gaussian mixture model for urban link travel time analysis. J. Transp. Syst. Eng. Inf. Technol. **16**(04), 178–184 (2016)
9. Jin, S., Luo, X., Ma, D.: Determining the breakpoints of fundamental diagrams. IEEE Intell. Transp. Syst. Mag. PP:1–1 (2018)

10. Susilawati, T.M.A., Somenahalli, S.: Travel time reliability and the bimodal travel time distribution for an arterial road. Road Transp. Res. **19**(4), 37–50 (2010)
11. Luo, X., Wang, D., Ma, D., et al.: Grouped travel time estimation in signalized arterials using point-to-point detectors. Transp. Res. **130**, 130–151 (2019)
12. Liu, Y., Guo, X., Zhou, R., et al.: Travel time prediction of main transit line based on multi-source data fusion. J. Transp. Syst. Eng. Inf. Technol. (2019)
13. Pu, W.J.: Analytic relationships between travel time reliability measures. Transp. Res. Rec. J. Transp. Res. Board **2254**, 122–130 (2012)
14. Lint, J.W.C.V., Zuylen, H.J.V.: Monitoring and predicting freeway travel time reliability: using width and skew of day-to-day travel time distribution. Transp. Res. Rec. J. Transp. Res. Board **1917**(1917), 54–62 (2005)
15. Shen, X., Zhou, Y., Jin, S., et al.: Spatiotemporal influence of land use and household properties on automobile travel demand. Transp. Res. Part D: Transp. Environ. **84**, 102359 (2020)
16. Taylor, M.A.P.: Fosgerau's travel time reliability ratio and the Burr distribution. Transp. Res. Part B Methodol. **97**, 50–63 (2017)

Optimal Vehicle Performance Parameters Selection for Electric Bus Routes

Jinhua Ji, Mingjie Hao, and Yiming Bie

Abstract Electric buses (EBs) have the characteristics of zero emission and low noise, which plays an important role in reducing air and noise pollution and improving air quality in metropolitan areas. Choosing the appropriate electric buses type to meet the passenger demand and own interests is an important practical problem faced by many bus companies, among many types of electric buses. This paper conducts an optimization study on the vehicle and charger performance parameters selection for electric bus routes, with the objectives of minimizing the total operating cost of the bus company, including the annual average electric bus purchase cost, annual average charging facilities cost, and annual charging cost of electric bus fleet, and considers the passenger travel demand and integrity of bus service. Finally, two real electric bus routes are taken as an example to validate the proposed method. Results show that the optimized scheme is more conducive to saving the operation cost of bus companies compared with the current scheme.

Keywords Electric bus · Performance parameters · Operating cost

1 Introduction

With the decrease of battery price and improvement of performance of electric buses (EBs), the electrification of public transit buses has become an inevitable trend. EBs have the characteristics of zero emission and low noise, which plays an important role in reducing air and noise pollution and improving air quality in metropolitan areas. According to a study by Bloomberg New Energy Finance Electric, the number of electric buses will reach 1.2 million by 2025, accounting for more than 47% of the world's urban bus fleet [1]. By the end of 2019, China has 324,500 electric buses, accounting for 46.8% of the national urban bus fleet, an increase of 27.47% compared with the previous year. Many developed countries have also set electrification goals, such as Paris, Los Angeles and Copenhagen will achieve only electric buses by 2025, 2030 and 2031, respectively [2, 3].

J. Ji · M. Hao · Y. Bie (✉)
Jilin University, Changchun 130022, Jilin, China

© The Author(s), under exclusive license to Springer Nature Singapore Pte Ltd. 2021
X. Qu et al. (eds.), *Smart Transportation Systems 2021*, Smart Innovation,
Systems and Technologies 231, https://doi.org/10.1007/978-981-16-2324-0_6

49

In the face of many types of electric buses, how to choose the appropriate bus type to meet the demand and own interests is an important practical problem faced by many bus companies. The large EB has a large body size, which can carry more passengers at the same time. In the actual scheduling process, it can reduce the bus fleet size and the number of charging facilities. However, the purchase price of large buses is relatively expensive, the recharging time required is longer, and the unit mileage power consumption is more affected by its own curb weight. On the contrary, the approved passenger of small electric buses is limited. To meet the travel needs of passengers, a larger fleet size and a larger number of charging facilities are usually required. But at the same time, small electric buses are cheaper to buy and consume less power per mile. Therefore, combined with the operating mileage requirements of urban bus routes, the construction conditions of charging facilities, the price and performance parameters of electric buses, this paper establishes a performance parameter selection optimization model of electric bus and charging facilities, so as to optimize bus fleet and charging facilities for electric bus routes.

There are few studies on the direct selection of performance parameters of EBs for specific routes. Chen et al. established a mathematical model to determine the optimal deployment of various charging facilities and electric bus fleet size, satisfying the charging demand [4]. An developed a stochastic integer programming model to jointly optimize charging station locations and bus fleet size under random bus charging demand, considering time-of-use electricity tariffs [5]. Yao et al. established an optimization model with the minimum annual total scheduling cost with explicit consideration of differences in driving range, recharging duration and energy consumption of EBs for multiple vehicle types [6].

In addition, Ceder proposed the deficit function method to solve the vehicle scheduling problem based on a given vehicle type, in which the categories are sorted in descending order of vehicle cost [7]. Hassold and Ceder developed a vehicle scheduling method based on the minimum cost network flow, which considers the substitution between vehicle types [8]. They developed a bus dwell time estimation method at a bus bay [9] and a Trial-and-error train fare design scheme for addressing boarding/alighting congestion at CBD stations [10]. But these researches are about the traditional fuel bus.

Some study on the performance parameters selection optimization of electric bus is reflected in the transition from fuel bus to pure electric bus, including the optimization method of pure electric bus fleet size, performance parameters, charging infrastructure and other related parameters. Pelletier et al. presented a fleet replacement problem which allows organizations to determine bus replacement plans in a cost-effective way, namely considering purchase costs, salvage revenues, operating costs, charging infrastructure investments, and demand charges [11]. Islama and Lownes investigated the complete course of fleet replacement by minimizing the Life Cycle Cost of owning and operating a fleet of buses and required infrastructures [12]. Rogge et al. provided a methodology for the cost-optimized planning of depot charging battery bus fleets and their corresponding charging infrastructure. The defined problem covered the scheduling of battery buses, the fleet composition, and the optimization of charging infrastructure in a joint process [13].

Following the present introduction, Sect. 2 introduces the methodology for developing the model and shows its structure. In Sect. 3, we display a numerical example based on two real electric bus routes. Finally, conclusions are given in Sect. 4.

2　Problem Description and Formulation

The objective of this section is to establish a performance parameter optimization selection model for an electric bus route based on the principle of buses adapting to the route. We first provide a general description of the bus route, EBs, and charging facilities. The length of electric bus route is l km and the daily operation time is t_w min. Passenger demand is even and known during the whole day operation time. All EBs are uniformly arranged to the depot far away from the route, and recharging is conducted with low electricity prices at night. According to the different rated charging power, the chargers can be divided into J types, and the rated power of type j charger is denoted as $G_j, j = 1, 2, ..., J$. Let i be the serial number of each EB type, $i = 1, 2, ..., I$, and the main parameters of type i EBs mainly include the rated battery capacity Q_i^{nom} and the approved passenger P_i.

This paper conducts an optimization study on the performance parameters selection of EBs and chargers for a bus route, with the objective of minimizing the total operating cost, including the annual average electric bus purchase cost, the annual average charging facilities cost, and the annual charging cost of electric bus fleet.

2.1　Total Operating Cost for an Electric Bus Route

We define EBs leave from the starting station, run to the terminal, and then run back to the starting station as a complete turnaround. To meet the need of passengers and ensure the normal turnover of the bus route, at least the number of type i EBs put into operation every day N_i^d can be calculated by the following Eq. (1).

$$N_i^d = \frac{P_{\max} \times t_{\text{turn}}}{60 P_i \times \rho} \tag{1}$$

where $P_{\max}$ is the maximum section passenger flow of the bus route, passenger/h; ρ is the busload factor, %; t_{turn} represents the turnaround time, including the inbound trip travel time, outbound trip travel time and waiting time at the terminal, min.

Since EBs need regular inspection and maintenance, similar to that of study [11], we assume that each EB will complete its daily service according to the vehicle scheduling plan within η_1 days every year, and the remaining time will be run by substitute EBs. Then, the number of type i EBs needed yearly for the route N_i^y and their annual average purchase cost C_i^p can be obtained by Eqs. (2) and (3), respectively.

$$N_i^y = N_i^d \times \left(1 + \frac{365 - \eta_1}{\eta_1}\right) \tag{2}$$

$$C_i^p = \frac{C_i^b \times N_i^y}{Y_i^b} \tag{3}$$

where C_i^b is the unit price of type i EB, RMB/vehicle; Y_i^b is the service life of type i EB, year.

The service life of EBs are usually set at 8 years. However, the battery depreciation rate and actual service life of EBs are uncertain for the influence of the initial state of charge (SOC), depth of discharge (DOD) [14], and temperature of discharging. Lam and Bauer [15] tested the effect of various stress factors on the battery cycle life, and a practical capacity fading empirical model was proposed as Eqs. (4) and (5).

$$\xi_i\left(SOC_{avg,i}^d, SOC_{dev,i}^d, T_d\right) = \sum_{d=1}^{\eta_2}\left(\left(k_1 SOC_{dev,i}^d \times e^{k_2 SOC_{avg,i}^d} + k_3 e^{k_4 SOC_{dev,i}^d}\right)\right.$$
$$\left. \times e^{\left(-\frac{E_a}{R}\left(\frac{1}{T_d} - \frac{1}{T_{ref}}\right)\right)}\right) \times Q_i^d \tag{4}$$

$$SOH_i = \left(1 - \frac{\xi_i}{Q_i^{nom}}\right) \times 100\% \tag{5}$$

where ξ_i is the total capacity fade of type i EB, kWh; η_2 is the running days of type i EB, day; $SOC_{avg,i}^d$ and $SOC_{dev,i}^d$ are the average SOC and the deviation from the average SOC of type i EB per day, which are used to quantify the effects of initial SOC and DOD; R is the gas constant with the value of 8.314, J/(mol·K); T_d is the temperature per day and T_{ref} is the reference temperature, both in Kelvin; E_a is the activation energy and it is 78.06 kJ/mol; k_1, k_2, k_3 and k_4 are the fitting constant and their value are -4.092×10^{-4}, -2.167, 1.408×10^{-5}, 6.130, respectively; Q_i^d is the energy consumption of type i EB per day, kWh; SOH_i is the state of health of type i EB, %.

According to the definition of battery SOC, the relationships between $SOC_{avg,i}^d$, $SOC_{dev,i}^d$ and Q_i^d should satisfy Eqs. (6) and (7) [16, 17].

$$SOC_{avg,i}^d = \lambda_1 - \frac{Q_i^d}{2Q_i^{nom}} \times 100\% \tag{6}$$

$$SOC_{dev,i}^d = \frac{Q_i^d}{2Q_i^{nom}} \times 100\% \tag{7}$$

where λ_1 is the battery SOC of each EB at the start operating time every day, %.

When $SOH_i = 80\%$ the battery life is over, and η_2 in Eq. (4) is the maximum number of days that the type i EB can operate under fixed service intensity, correspondingly. The actual service life of type i EB Y_i^b can be calculated by Eq. (8).

$$Y_i^b = \frac{\eta_2}{\eta_1} \tag{8}$$

The energy consumption rates of different EB types are significantly different, because the energy density, size, and quantity of loaded battery cell are different. The energy consumption rate increases significantly with the increase of the curb weight of EB. Therefore, we fit the energy consumption of type i EB per day Q_i^d as shown in Eq. (9).

$$Q_i^d = \alpha_i \frac{t_w l}{t_{turn}} + Q_i^{s,\max} \tag{9}$$

where α_i is the energy consumption rate of type i EBs, kWh/km; $Q_i^{s,\max}$ is the average energy consumption of auxiliary system of type i EBs per day, kWh.

The ratio of EBs to chargers is determined by the type of charging facilities and the recharging time required for the optimized type of EB. When the battery SOC of all EBs are recharged to λ_1 every night, the ratio of type i EBs to type j chargers $\theta_{i,j}$ can be presented as Eq. (10).

$$\theta_{i,j} = \left\lfloor \frac{60N_i^d}{n_j \times t_{total}} \times \frac{Q_i^d}{G_j} \right\rfloor + 1 \tag{10}$$

where n_j is the number of EBs that each type j charger can service simultaneously, vehicle; t_{total} is the time range for electric bus fleet to recharge at night, min. Q_i^d/G_j shows the recharging time needed for each EB at night, h. The symbol $[\cdot]$ is a rounding function.

The annual average charging facilities cost C_j^q and the annual charging cost of the electric bus fleet C_i^o can be calculated by Eqs. (11) and (12), respectively.

$$C_j^q = \frac{N_i^d \times C_j^c}{\theta_{i,j} \times Y_j^c} \tag{11}$$

$$C_i^o = N_i^d \times Q_i^d \times C_a \times \eta_1 \tag{12}$$

where C_j^c is the unit price of type j charger, including the construction and installation cost RMB/pile; Y_j^c is the service life of type j chargers, year; C_a is the unit price of overnight recharging, RMB/kWh.

2.2 Parameters Optimization Model

In the process of optimizing the performance parameters of electric bus fleets and charging facilities, the interests of bus companies should be taken into account, and the passenger travel demand and integrity of bus service should also be ensured, i.e., service interruption will not occur due to electricity depletion.

The objective function and constraints of the optimization model are combined and listed as follows:

$$\min Z = C_i^p + C_j^q + C_i^o \tag{13}$$

$$\lambda_1 - \frac{Q_i^d}{Q_i^{nom}} \geq \lambda_2 \tag{14}$$

$$\lambda_1 - \frac{Q_i^d}{0.8 Q_i^{nom}} \geq 0 \tag{15}$$

$$\theta_1 \leq \theta_{i,j} \leq \theta_2 \tag{16}$$

$$N_i^d, G_j \in \mathbb{Z}^+ \quad i = 1, 2, \ldots, I; \quad j = 1, 2, \ldots, J \tag{17}$$

Equation (13) demonstrates the objective functions of the performance parameters optimization model, and the calculation method of C_i^p, C_j^q and C_i^o are shown in Sect. 2.1. Constraint (14) is to ensure the service integrity of all electric buses during daily operation time and return to the depot safely by sufficient electric quantity. λ_2 is the safety threshold of battery SOC specified by experienced bus dispatchers. Constraint (15) means that with the decline of battery life, the battery capacity of the electric bus can still meet the scheduling requirements of the whole day operation time. Equation (16) is the constraint for the ratio of EBs to chargers, its upper and lower limits are determined by the area occupied by the depot and recharging time required by other electric bus routes at night. Equation (17) lists the value ranges of optimization variables.

The parameter optimization model established in this paper is an integer linear programming model, which is relatively simple. In the process of practical application, for an electric bus route, the calculation scale of the optimization model is very small, and it can be solved directly. The branch and bound method and other heuristic algorithms can be used to improve the calculation speed [18–21], when the routes need to be optimized at the same time are large and there are many types of vehicles and charging facilities available.

3 Numerical Example

Meihekou city has been vigorously promoting the application of electric bus. At present, the city has 10 electric bus routes, a total of 75 electric buses, the total length of 177 km electric bus routes and the daily operating mileage of 11,430 km. The city is located in 42.5 N 125.6 E, which belongs to the humid climate zone of the middle temperate zone. The lowest and highest environmental temperatures are— 32 and 35 °C respectively, and the annual average temperature is 5.5 °C. We take the circular electric bus routes 101 and 103 as an example to assess the proposed parameters optimization model. The route 101 is 23 km long, with the turnaround time of 75 min and the daily operational time of 720 min. The total length of route 103 is 17 km with the turnaround time of 51 min and the daily operational time of 680 min. The maximum section passenger flow of the bus routes 101 and 103 is 948 passenger/h and 836 passenger/h during peak hours, respectively. EBs with size of $10.5 \times 2.5 \times 3.2$ m^3 and battery capacity of 202.93 kWh (i.e. type 5 EB in Table 1) are running on the routes.

The city is equipped with integrated DC fast charger in the depot, the output power is 120 kw, each charging pile can serve two buses at the same time, and the service life is 8 years. The cost of each new charger is 55600 RMB. Considering the charging demand of other electric buses, the charging times that can be allocated to routes 101 and 103 are 170 min and 120 min, respectively. The charging price at night is 0.42 RMB/h. There are six common types of pure EBs available, and the

Table 1 Relevant parameters of six common types of pure electric buses

i	Size (m^3)	Total mass (kg)	Q_i^{nom} (kWh)	P_i (passenger)	α_i (kWh/km)	C_i^b (RMB/vehicle)
1	$8.05 \times 2.35 \times 3.15$	12,000	110.59	61	0.510	440,000
2	$8.5 \times 2.5 \times 3.2$	13,000	133.51	69	0.601	580,000
3	$8.5 \times 2.5 \times 3.2$	14,000	162.30	70	0.626	620,000
4	$10.5 \times 2.5 \times 3.2$	16,500	199.37	79	0.671	680,000
5	$10.5 \times 2.5 \times 3.2$	16,500	202.93	76	0.687	700,000
6	$12 \times 2.55 \times 3.2$	18,000	249.76	89	0.786	900,000

Table 2 The values of other important parameters in the model

Parameter	Meaning	Unit	Value
ρ	The busload factor	–	1
λ_1	Lower bound of battery SOC	%	100
λ_2	Upper bound of battery SOC	%	10
η_1	Service days every year	day	292
T_{ref}	The reference temperature	Kelvin	298.15
θ_1	The minimum ratio of EBs to chargers	–	2
θ_2	the maximum ratio of EBs to chargers	–	4

relevant parameters are shown in Table 1. The values of other important parameters in the model are shown in Table 2.

For the above two electric bus routes, the 5th and 6th types of EB are feasible solutions. The 1st, 2nd and 3rd types of EB are difficult to meet the operation needs in the first year of operation (SOH = 100%), while with the decline of battery life, the 4th type of EB is gradually difficult to ensure the integrity of its service trip, so it has to end the service ahead of time or replace the shorter route. The optimal solution of the two EB routes are the type 6 EB, and the minimum total operating costs of routes 101 and 103 are 2,417,809.7 RMB and 1,375,733.6 RMB respectively. The comparison between the optimized and the current scheme is shown in Table 3.

According to Table 3, it can be found that the EB size and battery capacity obtained by the optimization model are larger than those of the electric bus currently in use. If the type 6 EB is used in route 101, its annual average total operating cost is 2417809.7 RMB, including annual average EBs purchase cost 2,076,923.1 RMB, annual average charging facilities cost 315,145.9 RMB and annual charging cost 25,740.7 RMB. If the type 6 EB is used in route 103, the average annual total operating cost is 1,788,502.3 RMB, including average EBs purchase cost 1,568,965.5 RMB, annual average charging facilities cost 198,974.7 RMB and annual charging cost 20,562.1

Table 3 Comparison between the optimized and the current scheme

Parameter	Route 101		Route 103	
	Optimize scheme	Current scheme	Optimized scheme	Current scheme
i	6	5	6	5
N_i^d (vehicle)	14	16	8	10
N_i^y (vehicle)	18	20	10	13
Y_i^b (year)	7.8	5.9	7.7	5.8
C_i^p (RMB)	2,076,923.1	2,389,078.5	1,176,470.6	1,568,965.5
C_j^q (RMB)	315,145.9	317,273.8	181,199.3	198,974.7
C_i^o (RMB)	25,740.7	29,263.2	18,063.7	20,562.1
Z (RMB)	2,417,809.7	2,735,615.5	1,375,733.6	1,788,502.3

RMB. Compared with the current 5th type, routes 101 and 103 can save 317,805.8 RMB and 412,768.7 RMB, respectively.

The battery capacity of the 5th and 6th types of EBs can meet the demand of service integrity in the whole life cycle. From the price of a single EB, the type 5 EB is indeed cheaper than the type 6, which may be the reason why the bus companies choose the type 5 EBs. However, it is limited by the size and small approved number of passengers. To ensure the normal turnover of the route, the bus fleet size required by the route will increase. As shown in Table 3, the number of type 6 EBs needed annually for the routes 101 and 103 are 2 and 3 less than that in the current scheme. At the same time, the optimized departure intervals of routes 101 and 103 are 6 min and 7 min respectively, which are 1 min more than the current scheme, having little influence on the waiting time of passengers and the service satisfaction of urban public transport system.

It is worth noting that the type 5 EBs have been in deep DOD, and the battery life are significantly shortened. If routes 101 and 103 adopt the type 6 EBs, the battery life of each EB will be extended by 1.9 years. The decrease of fleet size and the extension of battery life are the reasons why the annual average purchase cost of the type 6 EBs is lower than that of the type 5 EBs.

Obviously, when the upper limit of maximum charging time of each route is fixed, the increase of bus fleet size will promote the ratio of EBs to chargers, which hoist the annual average cost of charging facilities. Affected by bus size and curb weight, the unit mileage power consumption of type 6 is greater than that of type 5. However, the optimized result requires a smaller bus fleet size, which makes the annual average charging cost not increase but decrease. Therefore, type 6 EBs are more economical than type 5 EBs for routes 101 and 103 in terms of the annual average EBs purchase cost, annual average charging facilities cost, and annual charging cost.

4 Conclusions

In this paper, an optimal vehicle performance parameters selection model of electric bus fleet and charging facilities is established to help bus companies choose among many types of buses. On the basis of meeting the passenger travel demand and integrity of bus service, we aim to maximize the benefits of bus companies, that is, to minimize the total operating costs of bus companies, including the annual average EBs purchase cost, annual average charging facilities cost and annual charging cost. Finally, two real electric bus routes in Meihekou city are taken as examples to verify the effectiveness of the optimization model. The results show that the optimized scheme is more economical than the current one in all terms of the annual average EBs purchase cost, annual average charging facilities cost, and annual charging cost. Routes 101 and 103 can save the total operating costs of 317,805.8 RMB and 412,768.7 RMB, respectively.

Acknowledgements This study was supported by the National Natural Science Foundation of China (No. 71771062), China Postdoctoral Science Foundation (No. 2019M661214 & 2020T130240), Graduate Innovation Fund of Jilin University (No. 101832020CX154), and Fundamental Research Funds for the Central Universities (No. 2020-JCXK-40).

References

1. Electric Buses Will Take Over Half the World Fleet by 2025, https://www.bloomberg.com/news/articles/2018-02-01/electricbuses-will-take-over-half-the-world-by-2025.pdf. Last accessed 2018/4/3
2. Electric buses in cities: Driving towards cleaner air and lower CO_2. On behalf of: Financing Sustainable Cities Initiative, C40 Cities, World Resources Institute, Citi Foundation, https://data.bloomberglp.com/professional/sites/24/2018/05/Electric-Buses-in-Cities-Report-BNEF-C40-Citi.Pdf. Last accessed 2019/5/20
3. ZeEUS eBus Report #2: An updated overview of electric buses in Europe. URL. https://zeeus.eu/uploads/publications/documents/zeeus-ebusreport-2.pdf. Last accessed 2019/5/1
4. Chen, Z., Yin, Y., Song, Z.: A cost-competitiveness analysis of charging infrastructure for electric bus operations. Transp. Res. Part C: Emerg. Technol. **93**, 351–366 (2018)
5. An, K.: Battery electric bus infrastructure planning under demand uncertainty. Transp. Res. Part C: Emerg. Technol. **111**, 572–587 (2020)
6. Yao, E., Liu, T., Lu, T., Yang, Y.: Optimization of electric vehicle scheduling with multiple vehicle types in public transport. Sustain. Urban Areas **52**, 101862 (2020)
7. Ceder, A.: Public-transport vehicle scheduling with multi vehicle type. Transp. Res. Part C: Emerg. Technol. **19**(3), 485–497 (2011)
8. Hassold, S., Ceder, A.: Public transport vehicle scheduling featuring multiple vehicle types. Transp. Res. Part B: Methodol. **67**, 129–143 (2014)
9. Meng, Q., Qu, X.: Bus dwell time estimation at a bus bay: a probabilistic approach. Transp. Res. Part C **36**, 61–71 (2013)
10. Wang, S., Zhang, W., Qu, X.: Trial-and-error train fare design scheme for addressing boarding/alighting congestion at CBD stations. Transp. Res. Part B **118**, 318–335 (2018)
11. Pelletier, S., Jabali, O., Mendoza, J.E., Laporte, G.: The electric bus fleet transition problem. Transp. Res. Part C: Emerg. Technol. **109**, 174–193 (2019)
12. Islam, A., Lownes, N.: When to go electric? A parallel bus fleet replacement study. Transp. Res. Part D: Transp. Environ. **72**, 299–311 (2019)
13. Rinaldi, M., Picarelli, E., D'Ariano, A., Viti, F.: Mixed-fleet single-terminal bus scheduling problem: modelling, solution scheme and potential applications. Omega **96** (2020)
14. Xu, Y., Zheng, Y., Yang, Y.: On the movement simulations of electric vehicles: a behavioral model-based approach. Appl. Energy **283**, 116356 (2021)
15. Lam, L., Bauer, P.: Practical capacity fading model for li-ion battery cells in electric vehicles. IEEE Trans. Power Electron. **28**(12), 5910–5918 (2013)
16. Zhang, L., Zeng, Z., Qu, X.: On the role of battery capacity fading mechanism in the lifecycle cost of electric bus fleet. IEEE Trans. Intell. Transp. Syst. (2020). https://doi.org/10.1109/TITS.2020.3014097
17. Qu, X., Yu, Y., Zhou, M., Lin, C.T., Wang, X.: Jointly dampening traffic oscillations and improving energy consumption with electric, connected and automated vehicles: a reinforcement learning based approach. Appl. Energy **257**, 114030 (2020)
18. Gao, K., Yang, Y., Li, A., Li, J., Yu, B.: Quantifying economic benefits from free-floating bike-sharing systems: a trip-level inference approach and city-scale analysis. Transp. Res. Part A: Policy Pract. **144**, 89–103 (2021)

19. Gao, K., Yang, Y., Sun, L., Qu, X.: Revealing psychological inertia in mode shift behavior and its quantitative influences on commuting trips. Transport. Res. F: Traffic Psychol. Behav. **71**, 272–287 (2020)
20. Bie, Y., Xiong, X., Yan, Y., Qu, X.: Dynamic headway control for high-frequency bus line based on speed guidance and intersection signal adjustment. Comput.-Aided Civ. Infrastruct. Eng. **35**(1), 4–25 (2020)
21. Wang, S., Wei, Z., Bie, Y., Wang, K., Diabat, A.: Mixed-integer second-order cone programming model for bus route clustering problem. Transp. Res. Part C: Emerg. Technol. **102**, 351–369 (2019)

Evaluation and Optimization of Driver's Training Methods in View of Public Awareness

Zhuoxin Sun, Wanqing Long, and Weiwei Qi

Abstract In order to improve the training level of drivers, assist drivers to adapt to the actual driving environment, and effectively prevent traffic accidents. This paper based on a questionnaire survey of drivers and applicants for motor vehicle driving licenses and made the investigation of status quo of driver's training, public cognition analysis and driver's character analysis. This paper set about the public's willingness to drive, investigated the public's general awareness of safety, and collected the data of satisfaction survey. Through by comparison the state of before and after training, it considered the problems existing in current driver's training methods, as well as the nodes that can be optimized. This paper restored the cognition of drivers in different scenes, and acquired the psychological perception and cognition of the crowd through before and after the training. Specifically, it was divided into stress value, emotional value, caution and attention of psychological characteristics, visual sense, auditory sense, tactile sense and health status of physiological characteristics, and behavioral characteristics in different scenes. This paper summarized the driver's psychological and physiological characteristics, established the path among the driver's basic attributes, psychological characteristics, psychological characteristics, and traffic accidents, and established the structural equation model (SEM) model of driver's characteristics including recessive factors. Based on this, this paper proposed a classified driver's training method based on public cognition. According to the characteristics of the investigated population, the basic attributes are gender, age and education level, and then the types are classified according to the analysis of personality characteristics, so as to transfer the driving training from general training to targeted training. According to the characteristics of different drivers to be trained, the training intensity is different. It can make drivers develop more driving skills, more sufficient driving experience, enough safety awareness and high-quality driving character in the trainee stage to deal with the real driving environment.

Z. Sun · W. Qi (✉)
School of Civil Engineering and Transportation, South China University of Technology, Guangzhou, China
e-mail: ctwwqi@scut.edu.cn

W. Long
School of Management, Wuhan Textile University, Wuhan, China

Keywords Driver's training · Traffic safety · Quantitative analysis · Structural equation model

1 Introduction

In recent years, China's car ownership has increased rapidly. According to the statistics of the Ministry of Public Security in 2019, the number of newly registered motor vehicles in 2019 reached 32.14 million, and the number of motor vehicles in the country reached 348 million. The increase in car ownership has led to frequent road traffic accidents and poses a huge challenge for road safety management. Studies show that human factors account for a large proportion of traffic accidents, and the driver's own safety consciousness is an important influencing factor, which is accounting for 27.6% [1]. In order to reduce traffic accidents, scholars have carried out a lot of research related to autonomous driving [2–4], but it still takes time to improve [5, 6]. At this stage, the driver's driving level has a significant impact on traffic safety. Previous studies have shown that a reasonable driver training program can help drivers develop good driving habits and safety awareness [7, 8], and has great potential in improving road traffic safety [9]. Pradhan pointed out that the proportion of young novice drivers in fatal and non-fatal car accidents is too high, and one of the main reasons is that compared with systematically trained drivers, their ability to quickly identify potential threats is weak [10].

At present, the qualification examination in china generally carries on the targeted training by the driving training institution. However, the current domestic training form has certain limitation, the driver initial education mainly is the transportation regulation, the basic driving technique and so on. In order to solve the limitations of domestic driving training, scholars have carried out a lot of research. Li proposed t puts forward an experiential teaching and training methods to provide various forms of experience education, which is using warning education site and 3D accident simulation to carry on the consciousness level prompt, and forms the strong warning education effect. [11]. Bedinger propose a training method that can be applied to different training institutions for task-level decomposition of operating vehicles. The core is safety, efficiency and environmental friendliness as the triple standard [12]. Chen developed a driving game for the lack of risk detection for drivers, results show that playing danger perception game is an effective way to improve the risk detection skills of novice drivers [13].

Generally speaking, the training of drivers at home and abroad is mainly led by training institutions. Although scholars at home and abroad have carried out a lot of research on driver training, the existing programs are not targeted and difficult to operate and promote. To solve this problem, this paper studies the physiological and psychological cognitive characteristics of drivers in different safety scenarios based on the structural equation model. The research results reflect the safety awareness level and safety cognitive characteristics of drivers, which can provide a good theoretical basis for the development of targeted driver training system.

2 Materials and Methods

2.1 Sample and Procedure

The actual survey took an online survey, taking legal drivers after professional training and the trainee drivers as the survey object, and making an online questionnaire based on the 'Questionnaire Star' platform. The questionnaire was divided into two stages. In the first stage, different people were surveyed without certain objects, and the total number of questionnaires was 107. In the second stage, the subjects were locked into drivers after professional training. The total number of questionnaires was 702.

Among them, A total of 72 drivers were interviewed in the top 107, Among them self-study self-examination for 3 people, Professional driving school training for 69 people.

Relevant studies have shown that the sample size of the structural equation model should be 5–10 times the number of question items. This questionnaire has 33 items including basic information. The minimum sample size should be 165, that is, the sample size of this survey is suitable.

2.2 Evaluation Content

In this paper, the questionnaire surveys the public's overall perception, and starts with the public's willingness to drive, investigates the general public's perception of safety, and examines driver characteristics for specific analysis. The unintuitive indicators of the training effect are reflected in each scene of actual manipulation, in order to understand the driver's characteristics in more detail and accurately, the questionnaire obtains its index from the driver's different scene cognition. At the same time, the psychological perception and understanding of the crowd before training were obtained. The specific refinement is (1) Psychological characteristics: stress value, emotional value, caution, attention, safety awareness and personal characteristic. (2) Physiological characteristics: vision, hearing, touch and health value. (3) Simulation of common accidents: failure to allow accidents, rear-end accidents, retrograde accidents, non-human accidents, reversing accidents, parking and braking.

The psychological characteristics of drivers are investigated from the stress of driving in the field, the uneasiness of meeting the car, the humble reaction of driving, the decision of wrong route, the driving psychology of dangerous situation, the adjustment of driving state comfort state and the psychological reaction in special situation. To obtain more real questionnaire data, among them, Quantifying the problem options into four-point indices for analysis, corresponding to Stress Values (ST), Caution (CA), Attention (AT), Safety Awareness (SA), and Personality Characteristic (PC). Personality traits were also quantified in four-point terms according to common driver personality traits: Anger (PC-A), Humility (PC-H), Impatience (PC-I), Courage (PC-C), and Mindful (PC-M), PC = H + C + M-A-I, the full score is 12.

The process of driving a vehicle is the process of continuous circulation in the stages of feeling, perception, judgment and manipulation. The physiological characteristics of the driver have a great influence on the whole process. Physiological characteristics mainly refer to the driver's routine health state, which consists of auditory sensitivity H, visual sensitivity V, tactile sensitivity T, the possibility of fatigue driving FD and the possibility of drunk driving DD.

The physiologic characteristics were quantified as follows: the interviewee said she was completely drunk, quantified as DD = 0; Fatigue driving is an extremely easy problem for many drivers to ignore, Hence the fatigue driving FD = average (habit values, A formula of 0.3 + 0.7) for mild, the interviewee stated that she would not choose to continue driving after being in a long driving state, and the habit value = 0. And occasionally continue driving in sleepy situations, A mild value = 1. And occasionally when she is feeling unwell, so the severity = 1, FD = 0.5(full score of 4); Always watch the front and rearview mirrors, V = 3; She is not sure if be aware of the abnormal sound outside, H = 2; Before driving, always determine in advance that the pedal seat is comfortable and tactile, T = 3; Physiological characteristics = H + V + T-DD-FD (12 points as full). The physiological characteristics of the interviewee are 7.5.

Again, the probability of an accident is quantified as follows: The probability of failure is zero; The probability of a rear-end accident is zero; The retrograde probability is 2; The probability of parking braking; Non-human causes and other accidents is quantified as 0; The probability of drunk driving fatigue driving accident is 0.25 + 0.3 * trailing + 0.3 * average (retrograde, non-human accidents, drunk driving, stop braking, other), The likelihood of an accident is 0.135 (a full score of 4) [14].

2.3 Establishment of Structural Equation

In order to better evaluate the overall characteristics of the driver before and after training, it is planned to use the structural equation model for exponential evaluation. The structural equation model is a new statistical analysis method that has appeared since the middle of the twentieth century. It is also known as the statistics of recent years. One of the three major advances in learning. Structural equation models include measurement models and structural models. The measurement model part finds the relationship between the observed index and the latent variable; while the structural model finds the relationship between the latent variable and the latent variable.

The whole idea is to use the structural equation model to analyze the relationship between variables, construct a theoretical model according to the research purpose, and then use the measured data to verify the rationality of the theoretical model. Then

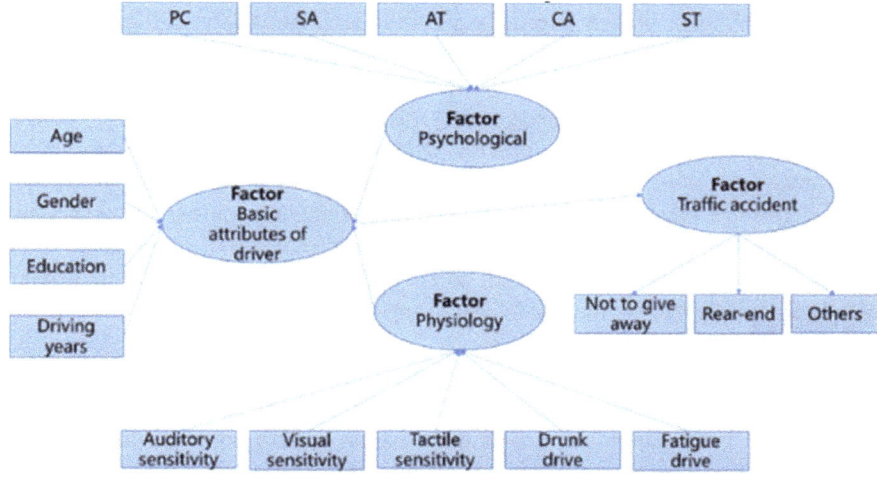

Fig. 1 sets up the driver characteristic SEM model

evaluate the model, and finally modify the model according to various evaluation indicators.

Therefore, this article continues to re-quantify the various indicators into variables with a full score of 1, thereby establishing a preliminary model based on the previous article. Since the basic attributes of the driver are affected by the psychological and physiological characteristics, the driver reflects the driving through a traffic accident. Skills proficiency and safety awareness. Establish the SEM model according to Fig. 1 and substitute the data.

Then compare according to the regression coefficients of the driver characteristic model: According to the standardized path coefficient value to judge the standard z and significance p, and check whether the influence relationship of the structure model is established. And use the calibration index to compare with the standard value to test the fitting effect of the model. The modeling calibration index in this paper is: chi-square freedom ratio, GFI, RMSEA, RMR, CFI, NFI, SRMR.

According to regression coefficients and calibration indicators, the relationship between each factor and its influence is further adjusted, and finally a structural equation model of driver characteristics with better fitting effect is established.

3 Results and Analysis

3.1 Descriptive Statistical Analysis

The current driving training is mainly self-study self-examination and professional driving school training, but self-study self-examination is only open in some cities in the country, most prospective drivers can choose the way is still professional driving school training. Professional driving school training is generally divided into theoretical knowledge training and field operation training. According to national statistics, women accounted for about 70% of the total number of cars, while men accounted for about 30% of the total (as of January 2019), and the results showed that women were more than twice as many as men. The age of the main car learners is also between 18 and 29 years old, indicating that young people have become the main driving training population, accounting for more than half of the share. This also means that driving training tends to be younger, that is, professional driving school training as long as the 18–29 age group of the car market, will seize nearly 80% of the share.

Among them, a total of 72 drivers were interviewed in the top 107, self-study and self-examination for 3 people, professional driving school training for 69 people. The actual driving manipulation training attention is 87.5 points (a full score of 100 points), The importance of theoretical training is only 58.33 points. Training satisfaction was 72.92, Safety awareness was 80.56, The average overall cowardice before training was 52.78, After training, fear was 38.54, Overall, the driving fear was reduced by 14.24 points. The total pre-diligence score was 39.70, the fear was 24.01, the overall reduction was 15.69 points. Interviewees only scored 58.33 points on theoretical, the overall safety awareness score was 80. And theoretical emphasis shows, Young people and middle-aged people pay less attention to theoretical training. People with less driving age also showed obvious dissatisfaction. The participants scored 87.5 points (100 points) for actual driving training. But after driving school training, driving proficiency is still low, the average of female proficiency is lower. People of different driving ages show dissatisfaction with driving training. It shows that the driving training for many years shows the characteristics of 'only do exercise but no learning'.

According to gender, age, driving age and education level, the psychological characteristics, physical characteristics and accident causes are clustered, and the differences between different categories are compared, and the maximum score is 12 points. Table 1 analyzes the clustering characteristics of different populations.

3.2 Structural Equation Model

The model is further modified, personality factor is removed, and emotional value factor is introduced to replace it.

Then, the evaluation index of the model is checked, except for the ratio of the degree of freedom of chi-square, the other main evaluation indexes all meet the requirements, and the fitting effect is more in line with the requirements, and the model diagram is more in line with the objective facts (Table 2).

Finally, the result of model is as Fig. 2.

Table 1 Comparison of Indicators

Category	Personality index	Health values
Male	6.56	10.88
Female	5.51	10.25
Average	6.44	10.81
18–29	5.61	10.07
30–39	6.67	11.00
40–49	6.63	10.99
Over 50	6.09	10.48
Average	6.44	10.81
Less than one year	4.50	8.47
One to three years	5.97	10.45
Three to five years	6.19	10.38
Five to seven years	6.62	11.11
Seven to ten years	6.47	10.61
More than 10 years	6.59	11.02
Average	6.44	10.81
Junior high school and below	6.65	10.89
Secondary or high school	6.62	11.04
Junior college	6.40	10.78
Bachelor degree	4.97	9.42
Master and above	4.54	10.64
Average	6.44	10.81

Table 2 Comparison of Final Driver Characteristics Model

Indicators	CMIN	GFI	RMSEA	RMR	CFI	NFI	NNFI	SRMR
Criteria	<3	>0.9	<0.10	<0.05	>0.9	>0.9	>0.9	<0.1
Value	4.314	0.958	0.066	0.002	0.962	0.951	0.924	0.077

3.3 Overview of Driving Training Methods

In this paper, a classified driving training method based on public cognition is proposed. Firstly, age, sex and education are used as basic attributes input, gender and age are used as basic characteristics to classify, and the proportion of teaching methods is determined by academic qualifications (Fig. 3).

Then according to the analysis of personality characteristics, according to personal characteristics subdivision, using the method of personality test, can be through questionnaire or question-and-answer way to carry out personality index statistics,

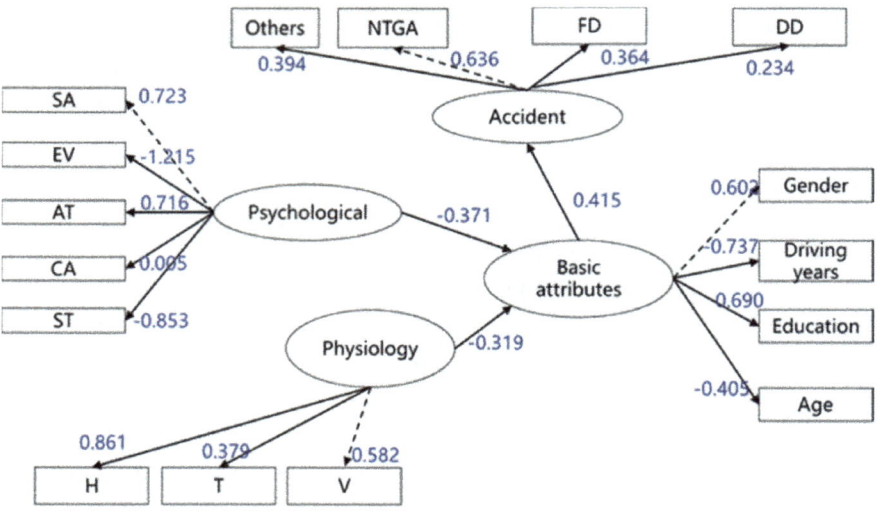

Fig. 2 Results of final driver characteristics model

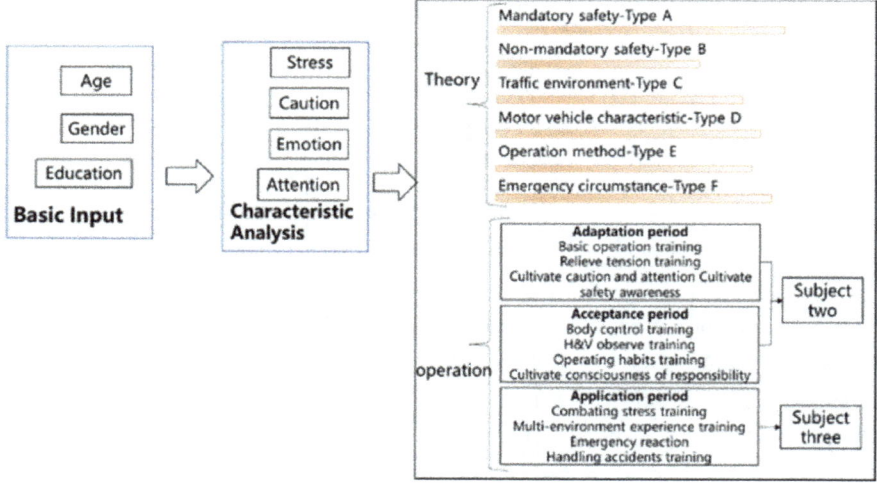

Fig. 3 Results of final driver characteristics model

at the same time set qualified line, according to traffic accident possibility statistical analysis, the probability of one-year and three-year driving is much lower than that of drivers.

The target values of each attribute are as follows: pressure value target value is set to intermediate value 2.0, caution target value is set to 3.0, focus target value is set to 2.5, safety consciousness is set to target value 4.0, emotion value is set to target value 1.0, character target value is set to target value 6.0.

Each type of training consists of theoretical knowledge training and practical training. In which theoretical knowledge training from general training to targeted training, Classification of theoretical knowledge as mandatory safe driving theory (category A), Non-mandatory safety theory (category B), Traffic environment theory (category C), Motor vehicle characteristics theory (category D), Normal manipulation method theory (category E) and special case contingency theory (category F), Assign school hours by category. According to subject 1 syllabus, Motor vehicle basic knowledge is recommended for 3 h. The recommended hours for laws, regulations and road traffic signals are 9 h, and this part is divided into A, B, C, D four categories. Subject 4 The syllabus provides for safe and civilized driving operation requirements, knowledge of safe driving under severe weather and complex road conditions, and methods for dealing with emergency situations such as tire burst. And knowledge of handling traffic accidents. But there is no set time, this part is divided into E, F categories. According to the rule of Ebbinghaus memory curve, this paper calculates the daily concentrated study time of one or two hours, a total of 15 days of study time.

After fully adapting to the real driving environment, we begin to carry out the multi-environment experience training of continuous driving, including highway (including tunnel, continuous downhill, etc.), urban road (including crowded meeting car, reasonable avoidance, etc.), rural road (including mountain road, bend, etc.) and so on. If the conditions do not allow the use of simulated driver experience teaching method for experience. At the same time, accident stress training and accident management training. Through the scene simulation before the real accident and the scene restoration after the accident, the students are trained to make subconscious response and action. Through the above training, students can get more driving experience training in a short time, which is beneficial to the improvement of the overall level of novice drivers.

4 Discussion

Through the investigation and design of this period, the following work has been completed: first, the characteristics of driver training methods at home and abroad are analyzed, the existing driving training methods are investigated and evaluated, and the existing problems of driving training are put forward. Second, study and analyze the characteristics of driver group. Thirdly, the model of driver's characteristic and accident-related structural equation is established by quantifying index. According to the structural equation model and group characteristics, a new idea is put forward and improved on the basis of the existing training methods, and a classified driving training method based on public cognition is put forward. The effect of the new training method is evaluated and analyzed through a small-scale investigation.

In the subsequent development period also encountered with a lot of problems. Due to the influence of the epidemic situation, the questionnaire collection adopts the method of online distribution, which leads to the screening of a large number of people with lower education, and the distribution scope mainly depends on the snowball distribution mode in the circle of friends. The randomness of the crowd was also affected.

However, in the future questionnaire survey, we will use the combination of offline access and online questionnaire as far as possible to ensure the average and rich collection of samples in each layer, and at the same time make the scene more diversified. Besides, we should continue to optimize the SEM model of driver's characteristics, introduce more influence factors, and improve the reliability and validity of reliability. In the optimization of driving training methods, there is still a lot of work to continue, in the subsequent research will also focus on the refinement of the method, and the effect of a larger and more accurate quantitative analysis.

Acknowledgements This work was sponsored by the Guangdong University Student Science and Technology Innovation Cultivation Special Fund Project of 2019 (No. pdjh2019b0035), and the Science and Technology Project of Guangdong Province (No. 2020A1414010148).

References

1. Chen, S.: Study on the application of AHP method in the cause analysis of urban road traffic accidents. Transp. World **5**, 129–131 (2008)
2. Wu, J., Kulcsár, B., Ahn, S., Qu, X.: Emergency vehicle lane pre-clearing: from microscopic cooperation to routing decision making. Transp. Res. Part B: Methodol. **141**, 223–239 (2020)
3. Zhao, X., Wang, Z., Xu, Z., Wang, Y., Li, X., Qu, X.: Field experiments on longitudinal characteristics of human driver behavior following an autonomous vehicle. Transp. Res. Part C: Emerg. Technol. **114**, 205–224 (2020)
4. Zhou, M., Yu, Y., Qu, X.: Development of an efficient driving strategy for connected and automated vehicles at signalized intersections: a reinforcement learning approach. IEEE Trans. Intell. Transp. Syst. **21**(1), 433–443 (2020)
5. Gao, K., Yang, Y., Li, A., Li, J., Yu, B.: Quantifying economic benefits from free-floating bike-sharing systems: a trip-level inference approach and city-scale analysis. Transp. Res. Part A: Policy Pract. **144**, 89–103 (2021)
6. Gao, K., Yang, Y., Sun, L., Qu, X.: Revealing psychological inertia in mode shift behavior and its quantitative influences on commuting trips. Transport. Res. F: Traffic Psychol. Behav. **71**, 272–287 (2020)
7. Horswill, M.S., Hill, A., Silapurem, L., Watson, M.O.: A thousand years of crash experience in three hours: An online hazard perception training course for drivers. Accid. Anal. Prev. **152**, 105969 (2021)
8. Wang, Y.C., Foss, R.D., O'Brien, N.P., Goodwin, A.H., Harrell, S.: Effects of an advanced driver training program on young traffic offenders' subsequent crash experience. Saf. Sci. **130**, 104891 (2020)
9. Endriulaitienė, A., Šeibokaitė, L., Markšaitytė, R., Slavinskienė, J., Arlauskienė, R.: Changes in beliefs during driver training and their association with risky driving. Accid. Anal. Prev. **144**, 105583 (2020)
10. Pradhan, A.K., Fisher, D.L., Pollatsek, A., et al.: Field evaluation of a risk awareness and perception training program for younger drivers. Proc. Hum. Factors Ergon. Soc. Ann. Meet. **50**(22), 2388–2391 (2006)
11. Li, F.: On the Role of Experiential Education in Driver Safety Training in Operating Vehicles Heilongjiang Science and Technology Information, vol. 24, pp. 281–281 (2016)

12. Bedinger, M., Walker, G.H., Piecyk, M., et al.: A hierarchical task analysis of commercial distribution driving in the UK. Procedia Manufacturing **3**, 2862–2866 (2015)
13. Chen, N., Mourant, R.R.: A hazard perception training game for novice drivers. In: International Conference on Information Science and Digital Content Technology, pp. 28–31. IEEE (2012)
14. Bianchi Piccinini, G., Engstr, M.J., Rgman B.J., et al.: Factors contributing to commercial vehicle rear-end conflicts in Chin: a study using on-board event data recorders. J. Saf. Res. **62**, 143–153 (2017)

Optimization of Pure Electric Bus Scheduling Based on Immune Optimization Algorithm

Lianjie Ruan, Xiaoni Hao, and Weiwei Qi

Abstract In the past two decades, more and more people have gathered in cities, which are facing serious traffic congestion and environmental pollution. Shenzhen has taken the lead in completing the upgrade from fuel buses to fully electric buses, proving that it is inevitable for urban public transportation companies to adopt pure electric buses. However, due to the limitations of the current level of technological development, electric buses are marked by short cruising distance and long charging time, and traditional fuel bus scheduling models are no longer feasible. Research on the application of pure electric bus dispatching is imminent and difficult. Depending on the characteristics of pure-electric bus dispatch, this paper constructs an optimization model of vehicle dispatching with minimum cost from the perspective of operating enterprises and designs a solution process based on immune optimization algorithm. Finally, four bus routes are selected for example analysis to obtain driving dispatching schemes with different battery capacities and charging characteristics. The example results show that with the increase of the cruising distance and charging speed, the minimum fleet size required for dispatch and the corresponding dispatch cost have all decreased, but when it increases to a certain value, the characteristics of the bus no longer affect the dispatch plan.

Keywords Pure electric bus · Operational scheduling · Immune optimization algorithm

1 Introduction

With the rapid development of social economic, and great improvement of the level of urbanization and motorization, traffic congestion and environmental pollution are becoming more and more prominent. According to the big traffic data of AutoNavi Map in 2019, the proportion of unobstructed roads in China is only 39%. In megacities, the main source of air pollution is no longer industrial emissions, but traffic

L. Ruan · X. Hao · W. Qi (✉)
School of Civil Engineering and Transportation, South China University of Technology, Guangzhou 510641, China
e-mail: ctwwqi@scut.edu.cn

© The Author(s), under exclusive license to Springer Nature Singapore Pte Ltd. 2021
X. Qu et al. (eds.), *Smart Transportation Systems 2021*, Smart Innovation,
Systems and Technologies 231, https://doi.org/10.1007/978-981-16-2324-0_8

pollution. Shifting private car travel to public transportation is currently the main strategy to reduce traffic congestion, energy consumption, and emissions. Kun Gao's research shows that when the convenience of public transportation is improved, it not only improves the quality of public transportation services, but also reduces the possibility of car users to use cars, thus reducing traffic congestion [1]. Under the dual pressure of traffic congestion and environmental pollution, zero-emission pure electric vehicles for public transportation have received more and more attention.

Compared with traditional fuel vehicles, the most significant difference between pure electric buses is power source. Due to the development of battery technology, pure electric bus vehicles cannot reach the same cruising range as fuel buses, and traditional bus dispatching models cannot be applied to pure electric buses. The dispatch and operation of pure electric buses must pay attention to battery restrictions, that is, mileage and charging restrictions must be considered. Therefore, it is necessary to conduct scheduling studies based on the characteristics of pure electric buses. Antti Lajunen studied the life cycle cost of pure electric buses with different charging methods and different routes through simulation analysis [2]. Yueru Xu studied the microscopic motion model of pure electric vehicles in congested traffic flow. Based on the unique dynamics and characteristics of the vehicle, more accurate energy consumption and vehicle mileage can be obtained [3]. Baojun Tang established a life cycle cost model for pure electric buses. And through sensitivity analysis, it was determined that the purchase cost is the key to the application cost of pure electric buses [4]. From the perspective of user benefits, Kun Gao proposed an analysis method for the economic benefits of a single trip. The economic benefits of each trip were quantitatively evaluated in terms of travel time and usage costs through a travel choice model [5]. Under the premise of a given number of public buses, Palma constructed a single-line departure frequency optimization method based on passengers' unique expectations of delay costs [6]. Xueqin Niu established a single-line bus frequency planning model based on passenger flow demand, with the goal of maximizing the weighted average of passenger satisfaction and enterprise satisfaction [7]. Yiming Bie developed a bus travel time prediction model under the dynamic control method, which makes the arrival time interval of bus vehicles more uniform and improves the reliability of bus operation dispatching services [8]. Xiaobo Qu has developed a vehicle-following model using reinforcement learning methods for future self-driving pure electric bus vehicles, which effectively reduces delays in bus operation and improves the quality of operation from the aspect of vehicle driving, which can accurately reduce the size of the fleet [9]. However, during low peak periods, single-line scheduling is always troubled by idle vehicle resources. Amar Oukil proposed a method based on combined column generation and preprocessing variable stabilization to solve the problem of multi-site vehicle scheduling [10]. Enjian Yao found that the total operating cost of the electric bus area plan could be reduced by 18.20% [11]. Jun Li analyzed the effect of different charging and discharging strategies on the minimum vehicle demand on a single line [12]. Le Zhang analyzed the battery capacity decay in the battery charging and discharging cycles, and extended the service life of the battery through different charging strategies, thereby reducing the life cycle cost of the bus fleet [13]. Panhathai Buasri established a scheduling model for the electric bus charging demand related to the total driving distance and energy consumption rate, and solved the scheduling results under two different charging measures: charging at night and charging during the day [14]. With the objective of minimizing the number of vehicles required and minimizing the total distance traveled, Ming Wen developed a mixed integer programming model with a large-scale adaptive neighborhood search heuristic algorithm for the electric bus scheduling problem and presented a partial charging scheduling problem [15]. Tao Liu uses a formula based on the deficit function theory and an equivalent dual-objective integer programming model to minimize the total number of electric buses and the total number of chargers required for pure electric bus dispatch [16]. Jing Wang considered the relationship between battery load and battery capacity degradation and established a scheduling model to minimize the cost of battery replacement [17]. Matthias Rogge optimized the fleet size and its corresponding charging infrastructure based on cost-effectiveness to minimize the total cost of the entire fleet [18]. Xindi Tang proposed static and dynamic electric bus deconfiguration strategies to cope with random urban traffic conditions [19].

Among the scheduling optimization models, there are not only studies on the optimization of the bus company's operational efficiency or the single objective of passenger transportation cost, but also studies on multi-objective optimization that integrates the interests of passengers, bus companies, society, etc. They provide a solid theoretical basis for the pure electric bus scheduling optimization model studied in this paper.

2   Model Establishment

2.1   Problem Description

A set of bus departure and destination stations and the departure tasks running between them are known. The collection of known departure task information constitutes the departure schedule. The detailed information of the departure task includes departure station, departure time, arrival destination and arrival time, etc. The problem of pure electric bus scheduling is to select the appropriate vehicle for the tasks on the schedule, that is, to select the appropriate departure task for each vehicle. The bus executes a series of orderly departure tasks, that is, the vehicle scheduling plan. The task of the scheduling optimization model is to arrange the vehicles more optimally for the departure tasks of the departure schedule. First of all, it is necessary to ensure that the scheduling plan meets operating requirements, such as mileage, charging time, etc. and secondly, it is necessary to optimize the value of the objective function as much as possible.

2.2 Model Assumptions

Although the actual conditions are fully considered in this study, the factors affecting the daily operation of pure electric buses are very complicated. In order to adapt the model to more scenarios, we need some reasonable assumptions to simplify the actual conditions. These assumptions are as follows:

1. The vehicles on all lines are identical, that is, the vehicles on all lines have the same characteristics and are allowed to run on any other line.
2. The available charging equipment is able to meet the charging demand, ensuring that charging posts can be found when charging is required and that there are no queues. The performance of charging power with time is always fixed, that is, the relationship between charging capacity and charging time is linear.
3. The bus operation scheduling is based on a 24-h cycle. If the destination of the last trip in the bus cycle is not the departure station of the first trip, it needs to add a trip to return it to the original station.

2.3 Model Expression

Table 1 provide an overview of the notation used in the model descriptions.

Table 1 Natation table

Variable	Definition
C_D	The purchase cost of a single bus
C_p	Operating cost per mileage of a single bus
C_w	The unit time cost of waiting for departure at the station during operation
D	Total number of vehicles required
L_k	Average daily mileage of the k-th bus
T_{wk}	Average daily waiting time of the k-th bus
a_i	a decision variable. If the task has been scheduled for execution by the vehicle, then $a_i=1$, otherwise it is 0
B_e^j	The remaining power after trip (j)
B^j	the amount of electricity required for safe operation of trip (j)
C	Charging rate
$d0$	The cruising distance
T_{Arrive}	Arrival time
T_{Depart}	Departure time
T_{Charge}	Charging time
$T_{Deadline}^j$	After task (j), the time required for the bus to go to the departure station of the next task without stopping

Here, we can build the following scheduling model.

$$MinC = C_D * D + C_P * \sum_{k=1}^{D} L_k + C_w * \sum_{k=1}^{D} T_{wk} \tag{1}$$

Subject to:

$$\sum_{i=1}^{n} a_i = n \tag{2}$$

$$B_e^j + T_{Charge}^j * \frac{C}{d0} > B^{j+1} \tag{3}$$

$$T_{Arrive}^j + T_{Charge}^j + T_{Deadline}^j \leq T_{Depart}^{j+1} \tag{4}$$

$$B^{j+1} * \frac{d0}{C} \leq T_{Charge} \leq \frac{1}{C} \tag{5}$$

$$T_{Depart}^{j+1} - T_{Arrive}^j - T_{Deadline}^j > T_{min} \tag{6}$$

$$\sum_{j=1}^{n} T_{Arrive}^j - T_{Depart}^j < T_{max} \tag{7}$$

Known parameters of the model: departure timetable, operating data (mileage, time spent, etc.), purchase unit price of bus, operating cost per unit mileage, waiting costs in stations, characteristic data of public transport vehicles (endurance range, charging rate).

3 Solving Algorithm

3.1 Immune Optimization Algorithm

Inspired by the biological immune system, the immune optimization algorithm applies the immune concept and theories to the genetic algorithm. Under the premise of maintaining the excellent characteristics of genetic algorithm, immune optimization algorithm attempts to selectively and purposefully use some characteristic information of the problem to be solved to suppress the degradation phenomenon in the optimization process, so as to obtain the global optimal solution.

3.2 Encoding

Encoding is the process of converting the actual problem to be solved into a mathematical form that can be solved directly by an algorithm. Generally, there are binary encoding, integer coding, etc. In this article, integer encoding is used. For example, the simple antibody [1 2 3 4 1 3 2 4] means that the first and fifth departure tasks are executed by vehicle 1, the second and seventh departure tasks are executed by vehicle 2, the third and sixth departures tasks are executed by vehicle 3, and the 4th and 8th departure tasks are executed by vehicle 4.

3.3 Initial Population

According to the model proposed in Sect. 2, the specific steps for generating n antibodies are as follows:

The trip tasks are sorted according to the departure time to form a task table, which mainly includes trip information, departure time, departure station, arrival time, arrival station, mileage, and other information.

Arrange a vehicle for the first task. The vehicle is required to be randomly selected from the departure station. Record: vehicle mileage (calculated based on the relationship between battery status and mileage), remaining vehicle mileage (vehicle mileage minus route mileage), and task number.

Assign vehicles for the remaining trips. and find the idle vehicles that can perform trip(i). The requirements: (1) The remaining cruising mileage plus the mileage added by charging minus the empty trip mileage is greater than the route mileage; (2) The arrival time of the last vehicle to complete the execution plus the time of the empty trip is less than the departure time of the next task. And the following principles are considered: (1) Prioritize allocation of vehicles that meet both conditions; (2) Within the constraints of the two conditions, if there are no idle buses that meet the requirements, new buses need to be added to execute this task.

The antibodies obtained according to the above rules are all feasible solutions. But in order to ensure the eligibility of the solution, it is also necessary to pass a test function. The core of this test function is to check the time and cruising range of the vehicle used by the antibody.

3.4 Selection, Crossover, Mutation

First of all, calculate fitness. The fitness represents the objective function value of the current antibody, which is expressed as the total cost of all trips.

Crossover reflects the genetic recombination in the evolution of biological immunity, and it is the main way to generate new individuals for offspring. Usually the crossover operation is to select the antibodies i1, i2 that need to be exchanged within the population through two random numbers, and then determine the crossover position j1 by a random number. However, due to the large number of constraints in the vehicle scheduling problem, there may be no effective solution after the crossover. Therefore, the method used in this paper is to first select the antibodies i1 and i2 that need to be exchanged by two random numbers. After finding the same planned point in both antibodies, the position of such a point is randomly selected, and then the exchange of antibodies is performed after this point. After the exchange, invalid individuals may still be generated, so the two antibodies obtained after the exchange need to be tested.

Mutation is to imitate the genetic mutation process in biological immune evolution, and it is an important way to produce new individuals. The mutant antibody may no longer meet the constraints, so the mutant antibody needs to be tested. If the requirements are met, the mutation is completed and the next antibody mutation can be performed. If it fails, continue until the maximum number of times is reached, forcing the mutation process of the next antibody.

The degree of optimization is related to the number of iterations and calculation time. As the number of iterations and calculation time increase, a better solution may appear. When the calculation time increases and the optimal objective function does not change, it can be considered that the optimal solution of the model under this parameter condition has been reached.

4 Case Analysis

4.1 Status of Bus Operation

In this paper, four bus lines (180, 195, 257 and 884) running in both directions in Guangzhou are selected for example analysis. After a detailed investigation and reasonable simplification of the operating data of these buses, the operating data is as Table 2.

After collating the departure timetable, a total of 512 departure tasks were obtained for a departure timetable running 16 h per day (6:00 am and 10:00 pm).

4.2 Parameter Calibration

The vehicle acquisition cost factor (C_D) is the unit price of the bus. The average price is taken as $1.5 million.

The driving cost conversion coefficient (C_p) is the unit operating cost of the bus. According to the research results of Bin Zhou, combined with the operating days of the bus vehicle (taken as 353 days) and the service life (8 years), it is taken as 1

Table 2 Bus operation data

Route No.	Direction	Length (km)	Departure interval (min)	Drive-time (min)	Empty trips drive-time (min)
257	Guangzhou Railway Station—Tianhe Coach Terminal	10	15	30	15
195	Lijiao Station—Guangzhoudong Railway Station	20	20	70	35
180	Lijiao Station—Guangzhou Railway Station	15	15	60	30
884	Tianhe Coach Terminal—Guangzhoudong Railway Station	5	10	20	10

yuan/km. Therefore, the average driving distance needs to be multiplied by 2824, and the driving cost conversion coefficient (C_p) is taken as 2824. The vehicle's charging cost factor can also be converted to a mileage cost based on the price of electricity.

The waiting time cost coefficient (C_w) is the cost factor for the time spent on standby at the station when the vehicle is in daytime operation. The coefficient is taken as 1.5 yuan/minute, which is calculated by combining the operating days and service life like the driving cost conversion coefficient (C_p).

In the pure electric bus dispatching model of this paper, battery capacity and charging rate affect the whole dispatching plan, so battery capacity and charging rate are used as optimization parameters to study their effects on the objective function.

The cruising distance (D0) can be a number between 100–300 km. The charging rate is an integer value of 1C-6C, that is, the charge time is taken as 60–10 min. The above values are substituted into the model for calculation.

4.3 Result Analysis

The results are calculated by MATLAB software. Taking the cruising distance of 100 km and charging speed of 5C as an example, the results of iterative solution of the model are as follows.

From the graph of iteration results of the algorithm (Fig. 1), it can be seen that the optimal fitness does not change after 150 iterations of the algorithm, and it can be considered that the optimal solution of the model under this parameter condition has been reached. The optimal vehicle scheduling solution under other parameters can be obtained by modifying the parameters.

In the same way, the dispatching plan of other numbered vehicles (24 vehicles in total) can be obtained (Fig. 2).

Fig. 1 Iterative results of the algorithm

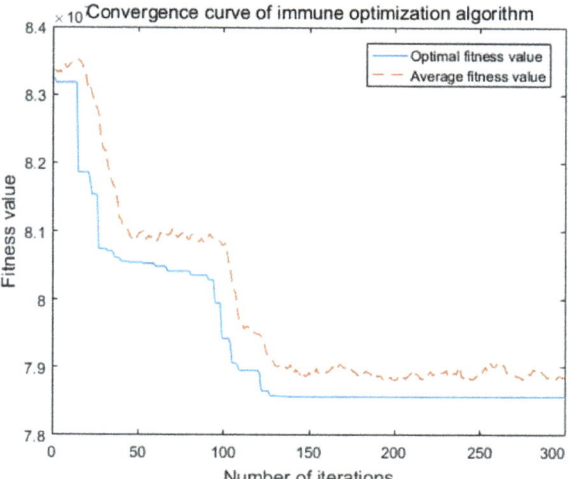

Fig. 2 Comparison between single-line scheduling and regional scheduling

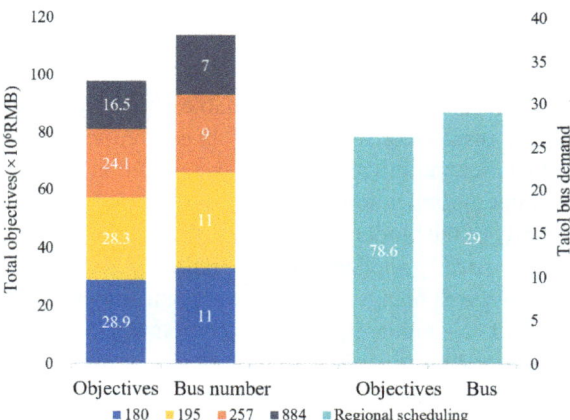

It can be seen that regional scheduling can effectively reduce fleet size and operating costs compared to single-line scheduling. Depending on the dispatching results obtained under different conditions, the optimization range is about 10% to 20%.

The biggest difference between pure electric bus scheduling and traditional fuel bus scheduling is the energy supply conditions. Today's electric buses are severely limited by their cruising distance and charging rate. These two factors are analyzed separately below.

As the cruising distance of pure electric buses gradually increases, the fleet size and operating costs gradually decrease. When the cruising distance reaches a certain level (cruising distance is greater than the maximum daily driving distance), the increase in cruising distance will no longer affect the scheduling plan, and thus the minimum operating cost and minimum fleet size will be determined (Fig. 3).

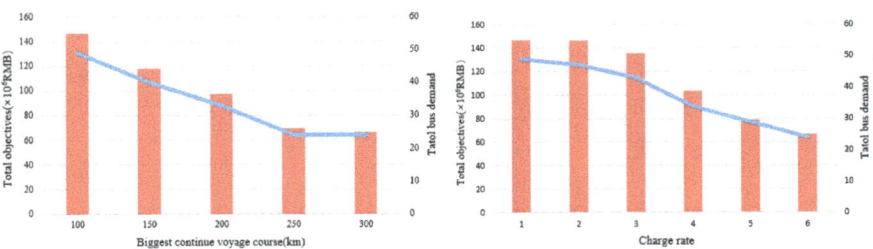

Fig. 3 The result of quantitative analysis

As the charging rate of pure electric buses increases, the operating cost and fleet size of the vehicle dispatching plan gradually decrease. When the cost of charging equipment is not considered, the dispatch cost and minimum fleet size of pure electric buses are inversely proportional to the charging rate.

5 Discussion

Pure electric buses are an inevitable choice for future development. In this paper, we use reasonable vehicle scheduling to reduce enterprise operation cost, establish a vehicle scheduling model under the condition of known departure schedule, and use immune optimization algorithm to solve the large-scale calculation problem. The validity of the model is verified in the actual bus lines in Guangzhou, and the method can be directly applied to the operation optimization of pure electric bus lines.

However, many problems were encountered in the study. Due to the influence of the epidemic and the equipment, a detailed analysis of different algorithms and multiple influencing factors is not presented in this paper. In addition, the detailed description of the algorithm processing and the comparison between different scheduling schemes are reduced due to the limitation of space. If necessary, If necessary, these can be studied in detail in future work.

Acknowledgements The study is supported by the National Natural Science Foundation of China (No. 52072131), the Science and Technology Project of Guangzhou City (No. 201804010466), the Fundamental Research Funds for the Central Universities (No. 2019MS120), the Science and Technology Project of Guangdong Province (No. 2020A1414010148), and the Key Research Projects of Universities in Guangdong Province (No. 2019KZDXM009).

References

1. Gao, K., Yang, Y., Li, A., Li, J., Yu, B.: Quantifying economic benefits from free-floating bike-sharing systems: A trip-level inference approach and city-scale analysis. Transp. Res. Part A: Policy Pract. **144**, 89–103 (2021)

2. Lajunen, A.: Lifecycle costs and charging requirements of electric buses with different charging methods. J. Clean. Prod. **172**, 56–67 (2018)
3. Xu, Y., Zheng, Y., Yang, Y.: On the movement simulations of electric vehicles: a behavioral model-based approach. Appl. Energy **283**. 116356 (2021)
4. Tang, B., Liu, J.: Economic analysis of the development of pure electric and hybrid electric buses in China. China Energy **35**, 37–41 (2013)
5. Gao, K., Yang, Y., Sun, L., Qu, X.: Revealing psychological inertia in mode shift behavior and its quantitative influences on commuting trips. Transport. Res. F: Traffic Psychol. Behav. **71**, 272–287 (2020)
6. de Palma, A., Lindsey, R.: Optimal timetables for public transportation. Transp. Res. Part B **35**, 789–813 (2001)
7. Niu, X., Chen, Q., Wang, W.: Optimization model of dispatching frequency of urban bus lines. J. Traffic Transp. Eng. **03**, 68–72 (2003)
8. Bie, Y., Xiong, X., Yan, Y., Qu, X.: Dynamic headway control for high-frequency bus line based on speed guidance and intersection signal adjustment. Comput.-Aided Civ. Infrastruct. Eng. **35**, 4–25 (2020)
9. Qu, X., Yu, Y., Zhou, M., Lin, C., Wang, X.: Jointly dampening traffic oscillations and improving energy consumption with electric, connected and automated vehicles: a reinforcement learning based approach. Appl. Energy **257**, 114030 (2020)
10. Oukil, A., Amor, H.B., Desrosiers, J., El Gueddari, H.: Stabilized column generation for highly degenerate multiple-depot vehicle scheduling problems. Comput. Oper. Res. **34**, 817–834 (2007)
11. Yao, E., Li, M., Liu, Y.: Electric bus area driving plan preparation considering charging constraints. J. South China Univ. Technol. (Nat. Sci. Ed.) **47**, 68–73 (2019)
12. Li, J., Tang, X., Zhao, C.: Dispatch optimization of pure electric buses based on charging strategy. J. Chongqing Jiaotong Univ. (Nat. Sci. Ed.) **34**, 107–112 (2015)
13. Zhang, L., Zeng, Z., Qu, X.: On the role of battery capacity fading mechanism in the lifecycle cost of electric bus fleet. IEEE Trans. Intell. Transp. Syst. 1–10 (2020)
14. Panhathai, B., Rathakarn S., Kingpaiboon B.: A simple electric bus schedule using energy demand. In: International Conference on Transportation and Civil Engineering (ICTCE'15), pp. 33–38 (2015)
15. Wen, M., Linde, E., Ropke, S., Mirchandani, P., Larsen, A.: An adaptive large neighborhood search heuristic for the electric vehicle scheduling problem. Comput. Oper. Res. **76**, 73–83 (2016)
16. Liu, T., Avi, C.A.: Battery-electric transit vehicle scheduling with optimal number of stationary chargers. Transp. Res. Part C **114**, 118–139 (2020)
17. Wang, J., Kang, L., Liu, Y.: Optimal scheduling for electric bus fleets based on dynamic programming approach by considering battery capacity fade. Renew. Sustain. Energy Rev. **130**, 109978 (2020)
18. Rogge, M., van der Hurk, E., Larsen, A., Sauer, D.U.: Electric bus fleet size and mix problem with optimization of charging infrastructure. Appl. Energy **211**, 282–295 (2018)
19. Tang, X., Lin, X., He, F.: Robust scheduling strategies of electric buses under stochastic traffic conditions. Transp. Res. Part C: Emerg. Technol. **105**, 163–182 (2019)

Optimal Design of Mixed Charging Station for Electric Transit with Joint Consideration of Normal Charging and Fast Charging

Le Zhang, Ziling Zeng, and Kun Gao

Abstract In this paper, we propose a bi-level model to optimize the design of mixed charging station deployed at the terminal station for electric transit. At the lower level, the service schedule and charging strategy of electric buses is optimized under the given design of mixed charging station. The lower-level model is a management optimization at the operational level, aiming at minimizing the total daily operational cost. The model is formulated as a mixed integer programing subject to limited charging facilities of multiple types. At the upper level, the design of charging station is optimized based upon the results obtained at the lower level. It is a kind of decision making at the tactical level, with the objective of minimizing the sum of operational cost of electric bus fleet (i.e., objective function of the lower-level model) and installation cost of charging infrastructure. We conducted numerical cases to validate the applicability of the proposed model and some managerial insights stemmed from the numerical case studies are discussed, which can help transit agencies design charging station scientifically.

Keywords Electric bus · Charging station · Vehicle scheduling · Bi-level model

1 Introduction

Electric transit is considered as the key to the world's clean transport future due to its high energy efficiency, zero emissions [1–5] and shareability [6–9]. Compared with diesel buses, battery-electric buses are able to improve energy efficiency by 50% and reduce greenhouse gas emissions by 98.36% [10]. During the past decade, the public transit is electrified step by step. For example, in the United State, the share of electric buses in the transit bus market increased rapidly from 2% in 2007 to nearly

L. Zhang (✉)
School of Economics and Management, Nanjing University of Science and Technology, Nanjing 210094, China
e-mail: le.zhang@njust.edu.cn

Z. Zeng · K. Gao
Architecture and Civil Engineering, Chalmers University of Technology, 41296 Goteborg, Sweden

20% in 2015 [11]; in Europe, the percentage of electric buses on the sales volumes of city buses is up to 10% by 2019 and this number rises up to around 20% in 2020. Undoubtedly, transit electrification is becoming an unstoppable trend.

Compared with diesel buses, the driving characteristics and refueling manner of electric buses are distinct. Specifically, battery-electric buses generally have a much shorter operational range than buses powered by other energy sources, resulting in users' "range anxiety". To ensure normal operations, the consumed electricity must be replenished by either battery swapping or battery recharging [12, 13]. Unfortunately, instead of mitigating this disadvantage, lack of sufficient charging facilities further aggravates it. However, if sufficient charging facilities are deployed, it may cause severe budget burden for transit system. Therefore, how to design the charging station at terminal, trading off between charging availability and limited budget, becomes an important issue in transit electrification. To this end, we aim at studying the optimal design of charging station deployed at the terminal station for electric transit in this paper. To be specific, we propose a bi-level model, where the lower-level model optimizes the scheduling of electric buses given the design of charging station, including the number of charging facilities under different charging modes (i.e., fast charging and normal charging); the upper-level model optimizes the design of mixed charging station given different electric bus scheduling optimized at the lower level. In the bi-level approach, the lower-level problem, i.e., optimal scheduling of electric buses, is the key and difficult part of the work. Therefore, we next present the relevant studies in the realm of vehicle scheduling.

The bus scheduling problem consists of assigning buses to serve a series of timetabled trips with the objective of minimizing fleet size and/or operational costs. It is an extension of the well-known vehicle scheduling problem (VSP), which has been extensively studied in the literature [14, 15]. Generally speaking, VSP can be categorized into two groups: the single-depot vehicle-scheduling problem (SDVSP) [16–18] and the multiple-depot vehicle-scheduling problem (MDVSP) [19, 20]. Over the years, many varieties and extensions of VSP have been proposed to incorporate the real-word constraints and conditions, including vehicle scheduling problem with multiple vehicle types [21], vehicle scheduling problem with route constraints (VSP-RC) [22], the alternative fuel vehicle scheduling problem (AF-VSP) [12, 23] and electric vehicle scheduling problem (E-VSP) [24]. Among these varieties, VSP-RC, AF-VSP and E-VSP are strongly motivated by electric vehicles. To accounting for the specifics of electric vehicles, route duration or route distance [25] are constrained in VSP-RC. If vehicles are allowed to be refueled at given recharging stations to prolong the total distance, that is AF-VSP. However, traditional AF-VSP only considers full charging and the charging time is set as fixed. Specifically, the vehicle's fuel level is set to be full after visiting any recharging stations. For example, Li [12] incorporated vehicle waiting time at charging stations into the model, and the charging time was simplified as fixed by considering battery swapping. Later, E-VSP was proposed, where partial charging was allowed and the charging time was usually assumed to be a linear function of the charged amount.

In light of the above, in this paper, we would employ the latest study in electric vehicle scheduling to study the optimal design of mixed charging station deployed at the terminal, and the bi-level solution approach is adopted. Numerical case studies were conducted to validate the applicability of the proposed model. It reveals that it is a cost-efficient choice to deploy sufficient charging facilities at the terminal station as the unit cost of charging facilities per day is much lower than that of electric buses.

The rest of this paper is organized as follows. Section 2 presents the problem formulation, i.e., a bi-level model. The numerical cases are conducted in Sect. 3. Conclusions are summarized in Sect. 4.

2 Formulation

In this section, a single-terminal transit network is considered (see Fig. 1) to define the optimization problem of charging station design, where electric buses depart from the terminal station to operate a sequence of scheduled round-trips, denoted as set V. Charging facilities in mode $q \in Q \triangleq \{1, 2\}$ are deployed at terminal station with limited capacity C_q, where $q = 1$ indicates normal charging and $q = 2$ indicates fast charging. For each round-trip $i (\in V)$, the departure time s_i, travel time e_i and the consumption of battery level relative to battery capacity, m_i are predefined and deterministic. The objective of this problem is to minimize the total cost of transit agency, including bus acquisition cost, charging fee, maintenance cost of electric bus fleet and the cost incurred by the deployment of charging facilities. Therefore, the operators shall make decisions at both tactical and operational levels. To be specific, at the tactical level, the number of charging facilities in each type deployed at the terminal station should be optimized and the vector of decision variables at this level is denoted by $C \triangleq \{C_q | q \in Q\}$. At the operational level, the operators need to make decisions on how to operate electric buses, and the vector of decision variables at this level (i.e., service sequence and charging strategy) is denoted by X.

We next present the lower-level model (i.e., at the operational level) and upper-level model (i.e., at the tactical level) in Sects. 2.1 and Sect. 2.2, respectively.

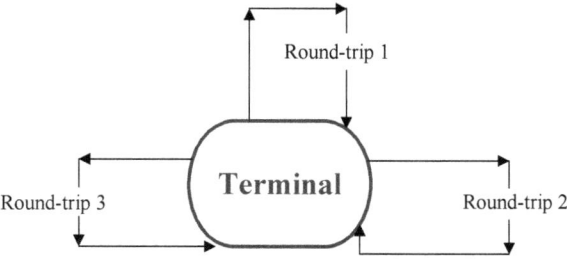

Fig. 1 Single-terminal transit network

2.1 Optimal Schedule of Electric Buses for Given Charging Station Design C

The objective of the lower-level model is to minimize total operational cost, including bus acquisition cost, charging fee and maintenance cost within one day, where the maintenance cost is mainly incurred by battery deterioration. In regard of this, we need to decide: (i) how to assign electric buses to serve a series of trips under the minimal battery level constraint; and (ii) how to optimize recharging schedules with limited charging facilities. It is worth to note that the total charging fee is constant in our model as it is related to the predefined trip service, which is fixed and independent of electric bus schedule. Therefore, the objective function is simplified as the sum of bus acquisition cost and maintenance cost. The time is discretized with the unit time as 10 min. In this model, we make the assumptions that the time for full charge in mode 1 (i.e., normal charging) and mode 2 (i.e., fast charging) are 2 h (i.e., 12 time steps) and 10 min (i.e., 1 time step) respectively.

The vector of decision variables at the operational level, X, is defined as follows:

$\delta_{ij} \in \{0, 1\}$: set to one if a bus serves round-trips i and j consecutively, where trip i begins earlier than trip j; and to zero otherwise, $i \in V \cup O, j \in V \cup D, i \neq j$; here, we use O and D simply to represent the original and destination depot respectively for notational convenience;

$\mu_{itq} \in \{0, 1\}$: set to one if bus begins to charge in mode q in the time step t after finishing round-trip i and to zero otherwise, $i \in V, t \in T, q \in Q$;

$\phi_{itq} \in \{0, 1\}$: set to one if bus is charged in mode q in the time step t after finishing round-trip i and to zero otherwise, $i \in V, t \in T, q \in Q$;

The lower-level model [P1] can be formulated as:

$$\min J_{[P1]} = \tilde{v} \cdot \sum_{i \in V} \delta_{Oi} + \sum_{i \in V} \sum_{t \in T} \sum_{q \in Q} \mu_{itq} d_q(SOC_i, 1) \tag{1a}$$

Subject to:

$$\sum_{i \in V \cup O} \delta_{ij} = 1, \quad \forall j \in V \tag{1b}$$

$$\sum_{j \in V \cup D} \delta_{ij} - \sum_{j \in V \cup O} \delta_{ji} = 0, \quad \forall i \in V \tag{1c}$$

$$\sum_{t \in T} \sum_{q \in Q} \mu_{itq} \leq 1, \quad \forall i \in V \tag{1d}$$

$$\left(1 - \sum_{t \in T} \sum_{q \in Q} \mu_{itq}\right) M + \sum_{t \in T} \sum_{q \in Q} \mu_{itq} \cdot t \geq s_i + e_i, \quad \forall i \in V \tag{1e}$$

$$\sum_{i \in V} \phi_{itq} \leq C_q, \quad \forall t \in T, q \in Q \tag{1f}$$

$$-(1 - \mu_{it1})M + 12 \leq \sum_{t'=t}^{t+11} \phi_{it'1} \leq (1 - \mu_{it1})M + 12, \quad \forall t \in T, i \in V \tag{1g}$$

$$-(1 - \mu_{it2})M + 1 \leq \phi_{it2} \leq (1 - \mu_{it2})M + 1, \quad \forall t \in T, i \in V \tag{1h}$$

$$1 - m_i - (1 - \delta_{0i})M \leq SOC_i \leq 1 - m_i + (1 - \delta_{0i})M, \quad \forall i \in V \tag{1i}$$

$$SOC_i \geq lb, \quad \forall i \in V \tag{1j}$$

$$SOC_j \leq SOC_i - m_j + (1 - \delta_{ij})M + M \sum_{t \in T} \sum_{q \in Q} \mu_{itq}, \quad \forall i, j \in V \tag{1k}$$

$$SOC_j \geq SOC_i - m_j - (1 - \delta_{ij})M - M \sum_{t \in T} \sum_{q \in Q} \mu_{itq}, \quad \forall i, j \in V \tag{1l}$$

$$SOC_j \leq 1 - m_j + (1 - \delta_{ij})M + M(1 - \sum_{t \in T} \sum_{q \in Q} \mu_{itq}), \quad \forall i, j \in V \tag{1m}$$

$$SOC_j \geq 1 - m_j - (1 - \delta_{ij})M - M(1 - \sum_{t \in T} \sum_{q \in Q} \mu_{itq}), \quad \forall i, j \in V \tag{1n}$$

$$d_q(SOC_i, 1) = \frac{2 \times \xi(SOC_i, 1) \times (1 - SOC_i)}{\chi} W_q, \quad \forall i \in V, q \in Q \tag{1o}$$

$$s_j \geq s_i + e_i - (1 - \delta_{ij})M - M \sum_{t \in T} \sum_{p \in Q} \mu_{itq}, \quad \forall i, j \in V \tag{1p}$$

$$s_j \geq \sum_{t \in T} \mu_{it1} \cdot t + 12 - (1 - \delta_{ij})M - M(1 - \sum_{t \in T} \mu_{it1}), \quad \forall i, j \in V \tag{1q}$$

$$s_j \geq \sum_{t \in T} \mu_{it2} \cdot t + 1 - (1 - \delta_{ij})M - M(1 - \sum_{t \in T} \mu_{it2}), \quad \forall i, j \in V \tag{1r}$$

In the above model, the objective function (1a) is to minimize the total operational cost over the operation hours of one day, including bus acquisition cost and maintenance cost, where $\tilde{v}$ denotes the unit acquisition cost of an electric bus per day, and d_q indicates the cost incurred by battery degradation with the state of charge (i.e., SOC) from SOC_i (i.e., the SOC of electric bus after finishing trip i) to 100% under charging mode q. Constraints (1b) guarantee that each round-trip is served exactly once. Constraints (1c) represent covering and flow conservation. Constraints (1d) state that after round-trip i, bus may start charging in a certain time step with a certain charging mode. Constraints (1e) ensure that the starting time of bus charging after round-trip i should be no earlier than the end time of round-trip i, where M is a sufficiently large number. Constraints (1f) guarantee the number of charging facilities used in each time step cannot exceed its capacity. Constraints (1g), (1h) state that 12 (1) time steps are occupied if normal (fast) charging operation is applied. Constraints (1i) indicate that the bus is fully charged when it departs from the original depot O. Constraints (1j) guarantee that SOC should be no smaller than a predefined lower bound lb to reduce range anxiety. Constraints (1k–1n) record the dynamic SOC of electric buses if $\delta_{ij} = 1$. Constraints (1o) define the function d_q where W indicates the battery acquisition cost, χ is the end-of-life related parameter. The term $\xi(SOC_i, 1)$ denotes the corresponding battery capacity fading rate, borrowed from Lam and Bauer [26]. Constraints (1p–1r) state the stating time of trip j should be no earlier than the ending time of trip i if $\delta_{ij} = 1$ and $\sum_{t \in T} \sum_{q \in Q} \mu_{itq} = 0$; and the stating time of trip j should be no earlier than the ending time of charging operation after trip i if $\delta_{ij} = 1$ and $\sum_{t \in T} \sum_{q \in Q} \mu_{itq} = 1$.

2.2 Optimal Design of Charging Station

The bi-level model can be formulated as follows:

The upper-level model [P2]:

$$\min_{C} J_{[P2]} = \sum_{q \in Q} A_q C_q + J_{[P1]}(X, C) \tag{2a}$$

subject to:

$$X = g(C) \tag{2b}$$

where A_q indicates the installation cost of charging facility in mode q amortized to one day, measured in \$/day; $g(\cdot)$ denotes the optimal lower-level solution for X under a given design of charging station C, which is found by solving model [P1]. $J_{[P1]}(X, C)$ indicates the daily operational cost under tactical decision C and operational decision X, which is consistent with the objective function of model [P1].

3 Numerical Cases

To validate our model, in this section we focus on a case study with the terminal denoted as terminal station A. Departing from terminal station A, five yet-to-be-electrified lines are studied. The lengths of the five lines are 11 km, 11.5 km, 12.6 km, 19.3 km and 16.8 km respectively. Their travel durations (terminal to terminal) are 80 min, 90 min, 105 min, 150 min and 130 min respectively. The timetables for these five lines are shown in Table 1. The technical parameters needed for this paper are obtained from Yutong ZK6850BEVG53, as specified in Table 2. The lower-level model is solved by the commercial solver SCIP. Here, for simplicity, we only consider one type of charging facilities, i.e., normal charging.

We optimize the electric bus service and charging schedule under a range of charging station capacity: $C_1 \in [5, 25]$, as presented in Fig. 2. The figure shows that the optimal operational cost (the solid curve with circle markers) decreases as C_1 increases, until it reaches a threshold of 16, as marked by the vertical dashed curve on the right-hand side of Fig. 2. This threshold represents the maximum number of charging facilities needed for electric buses at terminal station A; i.e., any additional charging facilities would be redundant, and the optimal total cost would stay the same ($\$1.106 \times 10^3$).

Table 1 Timetables for selected bus lines, departing from terminal station A

Line 1	Line 2	Line 3	Line 4	Line 5
07:00:00	06:10:00	07:00:00	06:20:00	06:40:00
07:20:00	Every 20 min	07:30:00	Every 30 min	07:10:00
07:40:00	09:10:00	08:00:00	18:50:00	07:40:00
Every 20 min	Every 30 min	Every 20 min		Every 10 min
10:40:00	15:10:00	14:20:00		18:20:00
Every 30 min	Every 20 min	14:45:00		18:50:00
16:40:00	19:10:00	15:15:00		19:20:00
Every 20 min		Every 20 min		19:50:00
19:20:00		20:35:00		

Table 2 Parameter definitions and values

Parameter	Notation	Value	Unit
Lower bound of battery level	lb	20	%
Unit acquisition cost of a electric bus (without battery) per day	$\tilde{v}$	16.5	$/day
Battery acquisition cost	W	28,000	$
Unit cost of a normal charging facility per day	A_1	5	$/day
End-of-life related threshold	χ	0.2	-

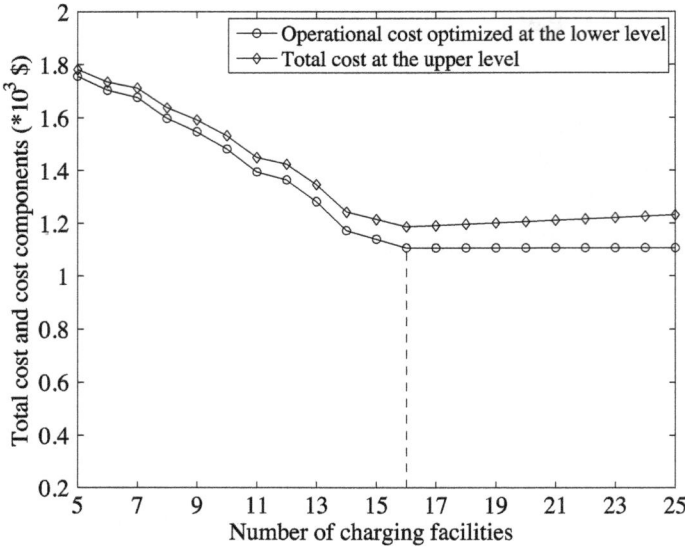

Fig. 2 Effects of the number of charging facilities on the optimal costs

The diamond-marked solid curve represents the corresponding total cost at the upper level. We note that, the optimal total cost decreases as C_1 increases, until it reaches to 16. After that, the optimal total cost increases as C_1 increases. This is because any more charging facilities would be redundant and the optimal operational cost at the lower level would not reduce any more. Therefore, for the studied transit network, it is optimal to deploy 16 charging facilities at the terminal station A. Further investigation reveals that, since the unit cost of charging facilities per day is relatively low as compared with that of electric buses, it is cost-efficient to deploy the charging facilities as more as needed so that the total cost of transit agency can be saved by reducing fleet size.

4   Conclusions

In this paper, we present a new mathematical formulation aimed at optimizing the design of charging station deployed at the terminal station for electric transit. To this end, a bi-level model is built with full consideration of the decision makings at both tactical and operational levels. Specifically, the lower-level model optimizes the scheduling of electric buses given the design of charging station, including the number of charging facilities under different charging mode (i.e., fast charging and normal charging), while the upper-level model optimizes the design of charging station given different electric bus scheduling optimized at the lower level.

In the future work we plan to explore more realistic scenarios where the "full-charging" assumption is relaxed, i.e., partial charging among electric buses is allowed and the charging time depends on the amount of energy to be replenished following the non-linear charging profile. Meanwhile, we would also plan to extend the transit network to the one with multiple terminals; explore more efficient solution approach with high quality, instead of using the commercial solvers; consider agency budget for the installation of charging infrastructure; and consider users' psychological inertia [27].

References

1. Lajunen, A.: Energy consumption and cost-benefit analysis of hybrid and electric city buses. Transp. Res. Part C: Emerg. Technol. **38**, 1–15 (2014)
2. Jin, S., Qu, X., Zhou, D., Xu, C., Ma, D., Wang, D.: Estimating cycleway capacity and bicycle equivalent unit for electric bicycles. Transp. Res. Part A: Policy Pract. **77**, 225–248 (2015)
3. Xu, Y., Zheng, Y., Yang, Y.: On the movement simulations of electric vehicles: a behavioral model-based approach. Appl. Energy **283**, 116356 (2021)
4. Qu, X., Yu, Y., Zhou, M., Lin, C.T., Wang, X.: Jointly dampening traffic oscillations and improving energy consumption with electric, connected and automated vehicles: a reinforcement learning based approach. Appl. Energy **257**, 114030 (2020)
5. Zhang, L., Zeng, Z., Qu, X.: On the role of battery capacity fading mechanism in the lifecycle cost of electric bus fleet. IEEE Trans. Intell. Transp. Syst. (2020). https://doi.org/10.1109/TITS.2020.3014097
6. Gao, K., Yang, Y., Li, A., Li, J., Yu, B.: Quantifying economic benefits from free-floating bike-sharing systems: A trip-level inference approach and city-scale analysis. Transp. Res. Part A: Policy Pract. **144**, 89–103 (2021)
7. Bie, Y., Xiong, X., Yan, Y., Qu, X.: Dynamic headway control for high-frequency bus line based on speed guidance and intersection signal adjustment. Comput.-Aided Civ. Infrastruct. Eng. **35**(1), 4–25 (2020)
8. Meng, Q., Qu, X.: Bus dwell time estimation at a bus bay: a probabilistic approach. Transp. Res. Part C **36**, 61–71 (2013)
9. Wang, S., Zhang, W., Qu, X.: Trial-and-error train fare design scheme for addressing boarding/alighting congestion at CBD stations. Transp. Res. Part B **118**, 318–335 (2018)
10. Mahmoud, M., Garnett, R., Ferguson, M., Kanaroglou, P.: Electric buses: a review of alternative powertrains. Renew. Sustain. Energy Rev. **62**, 673–684 (2016)
11. Neff, J., Dickens, M.: 2016 public transportation fact book. American Public Transportation Association, Washington, DC (2016)
12. Li, J.Q.: Transit bus scheduling with limited energy. Transp. Sci. **48**(4), 521–539 (2013)
13. Huang, Y., Zhou, Y.: An optimization framework for workplace charging strategies. Transp. Res. Part C: Emerg. Technol. **52**, 144–155 (2015)
14. Marković, N., Nair, R., Schonfeld, P., Miller-Hooks, E., Mohebbi, M.: Optimizing dial-a-ride services in Maryland: benefits of computerized routing and scheduling. Transp. Res. Part C: Emerg. Technol. **55**, 156–165 (2015)
15. Schöbel, A.: An eigenmodel for iterative line planning, timetabling and vehicle scheduling in public transportation. Transp. Res. Part C: Emerg. Technol. **74**, 348–365 (2017)
16. Paixão, J.P., Branco, I.M.: A quasi-assignment algorithm for bus scheduling. Networks **17**(3), 249–269 (1987)
17. Freling, R., Wagelmans, A.P., Paixão, J.M.P.: Models and algorithms for single-depot vehicle scheduling. Transp. Sci. **35**(2), 165–180 (2001)

18. Kang, L., Chen, S., Meng, Q.: Bus and driver scheduling with mealtime windows for a single public bus route. Transp. Res. Part C: Emerg. Technol. **101**, 145–160 (2019)
19. Dell'Amico, M., Fischetti, M., Toth, P.: Heuristic algorithms for the multiple depot vehicle scheduling problem. Manage. Sci. **39**(1), 115–125 (1993)
20. Kliewer, N., Mellouli, T., Suhl, L.: A time–space network based exact optimization model for multi-depot bus scheduling. Eur. J. Oper. Res. **175**(3), 1616–1627 (2006)
21. Ceder, A.A.: Public-transport vehicle scheduling with multi vehicle type. Transp. Res. Part C: Emerg. Technol. **19**(3), 485–497 (2011)
22. Bunte, S., Kliewer, N.: An overview on vehicle scheduling models. Public Transport **1**(4), 299–317 (2009)
23. Adler, J.D.: Routing and scheduling of electric and alternative-fuel vehicles (Doctoral dissertation, Arizona State University) (2014)
24. Wen, M., Linde, E., Ropke, S., Mirchandani, P., Larsen, A.: An adaptive large neighborhood search heuristic for the electric vehicle scheduling problem. Comput. Oper. Res. **76**, 73–83 (2016)
25. Haghani, A., Banihashemi, M.: Heuristic approaches for solving large-scale bus transit vehicle scheduling problem with route time constraints. Transp. Res. Part A: Policy Pract. **36**(4), 309–333 (2002)
26. Lam, L., Bauer, P.: Practical capacity fading model for Li-ion battery cells in electric vehicles. IEEE Trans. Power Electron. **28**(12), 5910–5918 (2012)
27. Gao, K., Yang, Y., Sun, L., Qu, X.: Revealing psychological inertia in mode shift behavior and its quantitative influences on commuting trips. Transport. Res. F: Traffic Psychol. Behav. **71**, 272–287 (2020)

Impact of Ambient Temperature on Electric Bus Energy Consumption in Cold Regions: Case Study of Meihekou City, China

Mingjie Hao, **Jinhua Ji**, and **Yiming Bie**

Abstract Electric buses are more environment-friendly due to their low noise and less air pollution. However, their electricity consumption on the route will change with operating conditions. According to field investigations, ambient temperature is one main contributing factor to energy consumption of an electric bus. When the temperature is very low, the energy consumption would increase significantly. The operational performance of the electric bus in cold regions should be examined carefully based on real world operation data. Thus, we choose Meihekou city, China which belongs to cold regions to collect ambient temperature and corresponding electricity consumption for six buses on a bus line. We gathered ambient temperature and corresponding electricity consumption of a trip in one day for six buses around a year to test the relationship between them. Pearson Correlation Coefficient is applied to verify the relevance of ambient temperature and electricity consumption. Results prove a negative correlation between them. After that, temperature and corresponding electricity consumption of the whole day for a year are studied. Ultimately, results illustrate electricity consumption variation is diverse during different seasons, and the largest electricity consumption is in winter. Results also show that when ambient temperature range drops from $[-2, 3]$ to $[-10, -2]$, change of electricity consumption is unstable and violent, which rises from 0.45 to 0.7 kWh/km. However, when ambient temperature ranges from -10 to -25.5 °C, the fluctuation of electricity consumption is small, which is dispersed between 0.6 and 0.7 kWh/km.

Keywords Cold regions · Ambient temperature · Electric buses · Electricity consumption

M. Hao · J. Ji · Y. Bie (✉)
School of Transportation, Jilin University, Changchun 130022, Jilin, China
e-mail: yimingbie@126.com

© The Author(s), under exclusive license to Springer Nature Singapore Pte Ltd. 2021   95
X. Qu et al. (eds.), *Smart Transportation Systems 2021*, Smart Innovation,
Systems and Technologies 231, https://doi.org/10.1007/978-981-16-2324-0_10

1 Introduction

Public transportation is more and more popular, which could alleviate traffic congestion and environmental pollution [1]. There are many studies about bus scheduling
i.e., bus route design, dynamic control, and timetable design [2–4]. The effect of
irrational psychological inertia in commuting mode shift behavior is studied [5].
Gao et al. provided an innovative approach that could quantify the benefits of using
bike-sharing systems at the trip level [6]. Meanwhile, Electric buses (EBs) have been
popular with the public in recent years. It is common for bus companies to replace
traditional fuel buses with EBs as they have the advantage of low-noise, low-polluting
and high driving stability, which could improve the comfort of passengers and the
public transportation attraction. Xu et al. [7] proposed a micro-traffic flow model
for EVs by considering their unique acceleration/deceleration characteristics. Zhang
et al. [8] proposed to constrain state of battery charge within a predefined range
and quantify its cost-effective feature through lifecycle cost analysis in view of the
practical battery capacity loss within charge and discharge cycling.

Studying the energy consumption for EBs is necessary [9–13], such as based on
Time-efficient stochastic model to predictive energy [14], the relationship between
energy consumption and well-to-wheels air pollutant emissions [15], electric vehicle-
routing problem [16] and the studies of energy demand uncertainty [17].

There are several studies about influential factors of electricity consumption [18,
19]. A forest-based stochastic model was established to consistently study the influence of environmental conditions, route characteristics and dynamic traffic conditions
[20]. Pamua and Pamuła [21] proposed a model which applied a reduced number
of readily acquired bus trip parameters: arrival times at the bus stops, map positions
of the bus stops and a parameter illustrating the trip conditions. Furthermore a deep
learning network was established for deriving the estimates of energy consumption
stop by stop of bus lines. Kivekäs et al. [22] developed a novel driving cycle syntheti-
zation method and this method was utilized to examine the effect of driving cycle
and passenger load changes on the energy consumption of EBs. Lajunen [23] introduced the cost–benefit analysis of hybrid and electric buses in fleet operation and
the analysis was based on energy consumption, which was collected from a large
number of simulations of different bus routes.

In addition, some studies regard ambient temperature as a influence factor of electricity consumption. Concretely, a two-dimensional grid stochastic energy consumption model based on a data-driven approach was established so as to distinguish the
influence of different ambient temperatures on the heterogeneous energy costs and
charging demand of autonomous electric vehicle fleets [24]. For the purpose of
evaluating the impact of ambient temperature on energy consumption and mileage,
and considering various reference driving cycles, a general quasi-steady backward-
looking model was proposed to estimate the energy consumption of electric vehicles
[25]. Based on GPS observations of 68 electric vehicles in Aichi Prefecture, Japan, an
energy consumption model was proposed and calibrated according to ordinary least
square regression and multi-level mixed effect linear regression. The results showed

the ambient temperature greatly affected energy efficiency by directly impacting output energy loss and interaction with vehicle auxiliary loads [26].

Based on previous studies, most of them work on many impact factors about electricity consumption. Some of them review the ambient temperature has an effect on electricity consumption. However, the existing literature does not consider that ambient temperature is taken as a single influence factor on electricity consumption of electric buses, using data from cold regions in China. Ambient temperature has a significant impact on electricity consumption of EBs, especially in cold regions where temperature in winter is extremely low. Therefore, examining electricity consumption change for EBs by ambient temperature in cold regions is of great importance. The purpose of this study is to discuss the electricity consumption variation of electric bus under different temperature conditions. Findings illustrate that electric bus manufacturers should pay attention to improve battery performance under the low temperature, moreover, the bus company should choose to purchase the electric bus that is suitable to cold regions and formulate different bus scheduling strategies as well as charging planning according to the change of electricity consumption by temperature variation.

The remainder of the paper is structured as follows: Sect. 2 is data description and processing of Meihekou city, China. Conclusions are drawn in Sect. 3.

2 Data Description and Processing

In order to study the relationship between electricity consumption and temperature, we collect the data for a year about bus electricity consumption from Meihekou city, Jilin province, China. The city is located in the north of China where belongs to cold regions and the ambient temperature is extraordinary low in winter. The data sample covers 6 electric buses marked as 1, 2, 3, 4, 5, 6, which are dispatched to run 108 line, 8.1 km. The charging planning is overnight charging, and not daytime charging.

(1)   Variation of electricity consumption under different ambient temperature

We collect electricity consumption and corresponding temperature of a trip in one day for 6 EBs and utilize Pearson Correlation Coefficient to verify the correlation between ambient temperature and electricity consumption, as shown in Table 1. It should be noted that T_i expresses the temperature of trip i and E_i indicates the electricity consumption of trip i.

Table 1 indicates relationships between ambient temperature and electricity consumption are significantly related for six buses. Relationship between ambient temperature and electricity consumption are negative correlation for six buses.

We take Bus 1 and Bus 3 as examples for showing the electricity consumption variation by ambient temperature, as shown in Figs. 1 and 2.

Figures 1 and 2 show electricity consumption vary significantly by ambient temperature. The temperature studied ranges from −25 to 35 °C and there is the same trend on two figures that electricity consumption decreases at first then maintains a

Table 1 Pearson correlations of six buses

			T_i	E_i
Bus 1	T_i	Pearson correlation	1	-0.684^{**}
		Sig. (2-tailed)		0.000
	E_i	Pearson correlation	-0.684^{**}	1
		Sig. (2-tailed)	0.000	
Bus 2	T_i	Pearson correlation	1	-0.315^{**}
		Sig. (2-tailed)		0.000
	E_i	Pearson correlation	-0.315^{**}	1
		Sig. (2-tailed)	0.000	
Bus 3	T_i	Pearson correlation	1	-0.594^{**}
		Sig. (2-tailed)		0.000
	E_i	Pearson correlation	-0.594^{**}	1
		Sig. (2-tailed)	0.000	
Bus 4	T_i	Pearson correlation	1	-0.430^{**}
		Sig. (2-tailed)		0.000
	E_i	Pearson correlation	-0.430^{**}	1
		Sig. (2-tailed)	0.000	
Bus 5	T_i	Pearson correlation	1	-0.148^{**}
		Sig. (2-tailed)		0.000
	E_i	Pearson correlation	-0.148^{**}	1
		Sig. (2-tailed)	0.000	
Bus 6	T_i	Pearson correlation	1	-0.544^{**}
		Sig. (2-tailed)		0.000
	E_i	Pearson correlation	-0.544^{**}	1
		Sig. (2-tailed)	0.000	

**Correlation is significant at the 0.01 level (2-tailed)

stable value with the increasing of temperature. Specifically, when temperature rises from -20 to -5 °C, electricity consumption per kilometer is more than 0.6 kWh/km. When temperature is higher than -5 °C, the change of electricity consumption per kilometer is gradually stable, mainly between 0.4 and 0.6 kWh/km. For different buses, electricity consumption is various. Maximum electricity consumption is 1 kWh/km for Bus 1 while 1.6 kWh/km for Bus 3. This is because many factors affect bus electricity consumption, such as driver's driving style, travel speed and so on, although Bus 1 and Bus 3 run the same line.

(2) Examining influence of temperature on electricity consumption

We collect ambient temperatures at different times in the whole day from January to December in 2020 of Meihekou city, moreover, the electricity consumption and driving mileage of six buses on 108 line in the whole day were collected, taking one

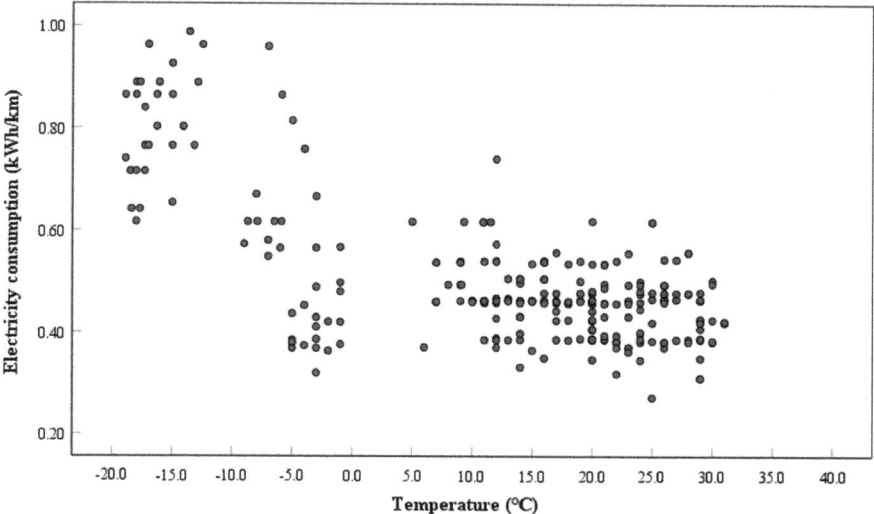

Fig. 1 Electricity consumption variation with ambient temperature for Bus 1

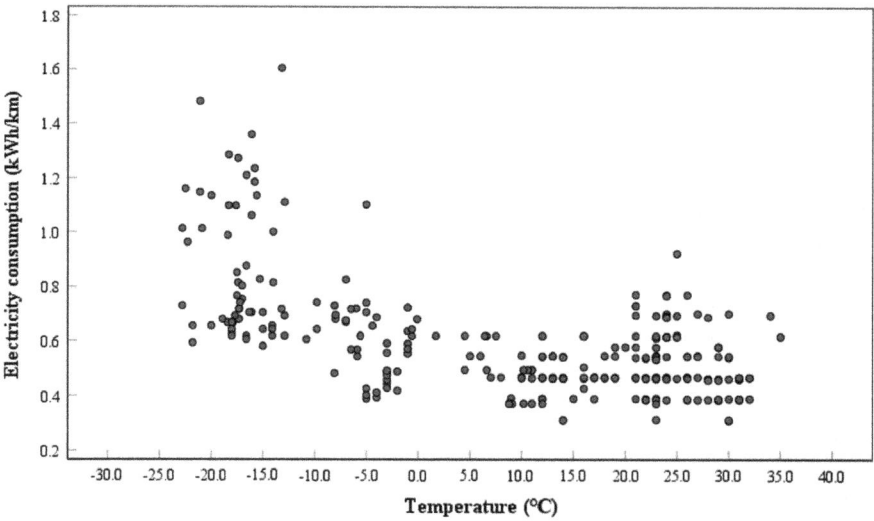

Fig. 2 Electricity consumption variation with ambient temperature for Bus 3

week as the sampling interval. Afterwards obtained data with large error is deleted to analyze the effective data. According to different months, one year is divided into four seasons, spring (from March to May), summer (from June to August), fall (from September to November) and winter (from December to February). We take Bus 1 and Bus 2 as a example to examine the change of electricity consumption by season.

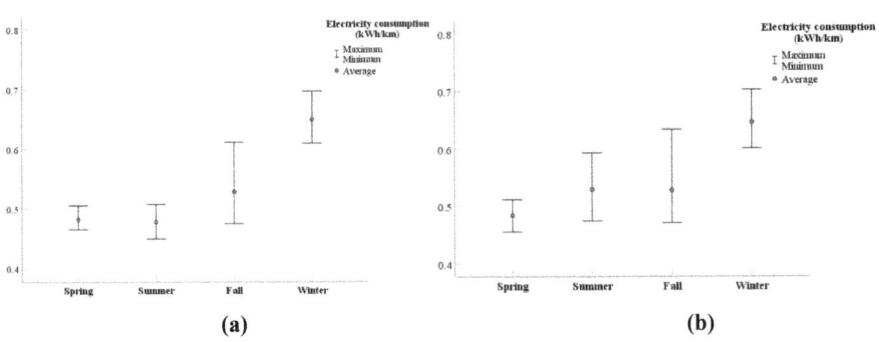

Fig. 3 Variation of electricity consumption by season for Bus 1 (**a**) and Bus 2 (**b**)

Figure 3a, b illustrate seasonal variation has a significant effect on electricity consumption. Vertical line denotes the range of electricity consumption in a certain season, and the top end of it represents the maximum electricity consumption while the bottom of it represents the minimum electricity consumption. The black dot at the vertical line indicates the average electricity consumption. The largest electricity consumption of two buses are both in winter, around 0.7 kWh/km, because the lowest temperature of cold regions is in winter. As temperature variation of fall is apparent, the change of electricity consumption in fall is more obvious than others. The minimum electricity consumption for Bus 1 and Bus 2 are both nearly 0.45 kWh/km.

According to gained ambient temperature, the daily temperature variation range is calculated, sorted based on the minimum temperature. Furthermore, the relationship between temperature and electricity consumption is represented by Fig. 4.

Figure 4 demonstrates the electricity consumption of all buses has an identical trend with the increasing of ambient temperature. On the whole, electricity consumption is reducing as ambient temperature is rising. The maximum electricity consumption for Bus 3 is 0.8 kWh/km and it is about 0.7 kWh/km for other buses and the minimum electricity consumption for six buses is nearly 0.45 kWh/km. Specifically, when ambient environment range is between [−25.5, −14.2] and [−10, −2], the fluctuation of electricity consumption is little. When the temperature range drops from [−2, 3] to [−10, −2], change of electricity consumption is unstable and violent, which rises from 0.45 to 0.7 kWh/km. Reasons are mainly by two aspects. On the one hand, bus drivers tend to choose to turn on the air conditioner to make passengers who are inside the bus feel warm, especially in cold regions where temperature is even lower than −25 °C in winter in order to ensure the temperature inside the bus is appropriate and improve the comfort of buses and passenger satisfaction under extremely cold weather. On the other hand, the performance of the battery will also be affected in extremely cold conditions, and the internal chemical reaction of the battery will be weakened by the low temperature, resulting in the increasing of electricity consumption.

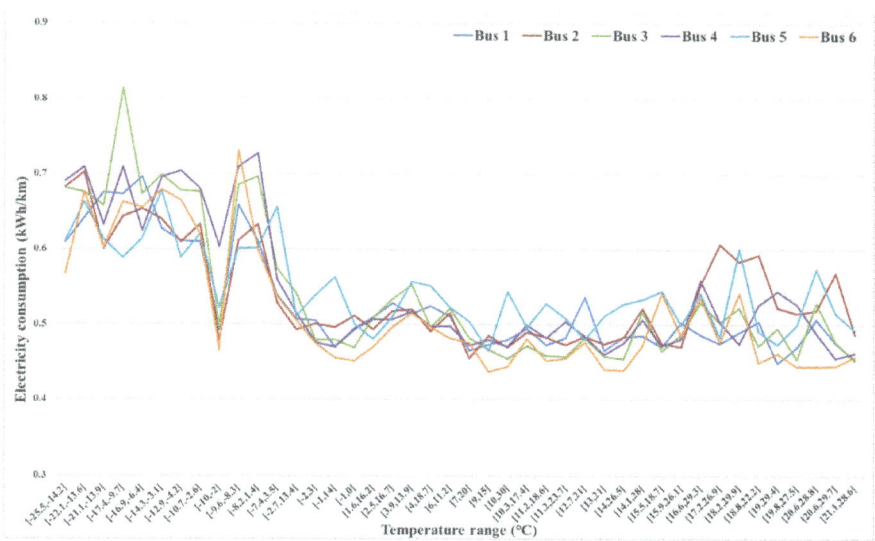

Fig. 4 Influence of temperature range on electricity consumption

Moreover, electricity consumption is nearly 0.5 kWh/km and the variation of that is stable when temperature range is from [−2, 3] to [15.9, 26.1], because ambient temperature is comfortable for passengers so that air conditioners do not need to be turned on under this circumstances, electricity consumption becomes less.

When ambient temperature range is from [15.9, 26.1] to [21, 28.6], electricity consumption increases slightly, but this change is less than the situation where the temperature range is between [−25.5, −14.2] and [−10, −2]. This is because the weather is so hot that air conditioner has to be turned on, however, the requirement for air-conditioning is not rather intense, because passengers could open windows to keep comfortable.

Furthermore, the variation of electricity consumption of Bus 3 and Bus 6 is obvious while that of others is inapparent with the change of ambient temperature. The major reason is that impact factors of electricity consumption for buses include not only ambient temperature but also the usage time of buses, the driving habits of drives and so on.

3 Conclusions

This paper analyzes the relationship between electricity consumption and ambient temperature. We utilize Pearson Correlation Coefficient to verify the correlation between electricity consumption and ambient temperature. The result proves a negative correlation between them. Curves that electricity consumption and ambient temperature of one trip for certain buses are drawn, it can be concluded that electricity consumption increases with the decreasing of ambient temperature. Meanwhile, seasonal variation could affect electricity consumption, and electricity consumption in winter reaches the maximum. In addition, we collect the temperature range and electricity consumption of the whole day, and it is found that the minimum electricity consumption is in winter. Besisdes, the result represents when ambient temperature changes from [−2, 3] to [15.9, 26.1], electricity consumption of six electric buses maintain nearly 0.5 kWh/km. When ambient temperature range is lower than [−2, 3] or higher than [15.9, 26.1], electricity consumption obviously increases and the maximum is 0.8 kWh/km.

In the future study, the influence of temperature on electricity consumption requires more buses on multiple routes to verify the validity of the results. Furthermore, more influence factors on electricity consumption should be studied.

Acknowledgements This study was supported by the National Natural Science Foundation of China (No. 71771062), China Postdoctoral Science Foundation (No. 2019M661214 and 2020T130240), Graduate Innovation Fund of Jilin University (No. 101832020CX154), and Fundamental Research Funds for the Central Universities (No. 2020-JCXK-40).

References

1. Wang, S., Zhang, W., Qu, X.: Trial-and-error train fare design scheme for addressing boarding/alighting congestion at CBD stations. Transp. Res. Part B **118**, 318–335 (2018)
2. Wang, S., Zhang, W., Bie, Y., Wang, K., Diabat, A.: Mixed-integer second-order cone programming model for bus route clustering problem. Transp. Res. Part C Emerg. Technol. **102**, 351–369 (2019)
3. Bie, Y., Xiong, X., Yan, Y., Qu, X.: Dynamic headway control for high-frequency bus line based on speed guidance and intersection signal adjustment. Comput.-Aided Civ. Infrastruct. Eng. **35**(1), 4–25 (2020)
4. Meng, Q., Qu, X.: Bus dwell time estimation at a bus bay: a probabilistic approach. Transp. Res. Part C **36**, 61–71 (2013)
5. Gao, K., Yang, Y., Sun, L., Qu, X.: Revealing psychological inertia in mode shift behavior and its quantitative influences on commuting trips. Transp. Res. Part F Traffic Psychol. Behav. **71**, 272–287 (2020)
6. Gao, K., Yang, Y., Li, A., Li, J., Yu, B.: Quantifying economic benefits from free-floating bike-sharing systems: a trip-level inference approach and city-scale analysis. Transp. Res. Part A Policy Pract. **144**, 89–103 (2021)
7. Xu, Y., Zheng, Y., Yang, Y.: On the movement simulations of electric vehicles: a behavioral model-based approach. Appl. Energy **283**, 116356 (2021)
8. Zhang, L., Zeng, Z., Qu, X.: On the role of battery capacity fading mechanism in the lifecycle cost of electric bus fleet. IEEE Trans. Intell. Transp. Syst. (2020). https://doi.org/10.1109/TITS.2020.3014097
9. Qu, X., Yu, Y., Zhou, M., Lin, C.T., Wang, X.: Jointly dampening traffic oscillations and improving energy consumption with electric, connected and automated vehicles: a reinforcement learning based approach. Appl. Energy **257**, 114030 (2020)
10. Al-Ogaili, A.S., Ramasamy, A., Hashim, T.J.T., Al-Masri, A.N., Hoon, Y., Jebur, M.N., Verayiah, R., Marsadek, M.: Estimation of the energy consumption of battery driven electric buses by integrating digital elevation and longitudinal dynamic models: Malaysia as a case study. Appl. Energy **280**, 115873 (2020)

11. Ma, X., Miao, R., Wu, X., Liu, X.: Examining influential factors on the energy consumption of electric and diesel buses: a data-driven analysis of large-scale public transit network in Beijing. Energy **216**, 119196 (2021)
12. Mahmoud, M., Garnett, R., Ferguson, M., Kanaroglou, P.: Electric buses: a review of alternative powertrains. Renew. Sustain. Energy Rev. **62**, 673–684 (2016)
13. Lajunen, A., Kivekäs, K., Baldi, F., Vepsäläinen, J., Tammi, K.: Different approaches to improve energy consumption of battery electric buses. In: 15th IEEE Vehicle Power and Propulsion Conference (VPPC), pp. 1–6. IEEE, Chicago, Illinois (2018)
14. Xie, S., Hu, X., Xin, Z., Li, L.: Time-efficient stochastic model predictive energy management for a plug-in hybrid electric bus with an adaptive reference state-of-charge advisory. IEEE Trans. Veh. Technol. **67**(7), 5671–5682 (2018)
15. He, X., Zhang, S., Ke, W., Zheng, Y., Zhou, B., Liang, X., Wu, Y.: Energy consumption and well-to-wheels air pollutant emissions of battery electric buses under complex operating conditions and implications on fleet electrification. J. Clean. Prod. **171**, 714–722 (2018)
16. Shao, S., Guan, W., Bi, J.: Electric vehicle-routing problem with charging demands and energy consumption. IET Intell. Transp. Syst. **12**(3), 202–212 (2017)
17. Vepsäläinen, J., Kivekäs, K., Otto, K., Lajunen, A., Tammi, K.: Development and validation of energy demand uncertainty model for electric city buses. Transp. Res. Part D Transp. Environ. **63**, 347–361 (2018)
18. Huda, N., Kaleg, S., Hapid, A., Kurnia, M.R., Budiman, A.C.: The influence of the regenerative braking on the overall energy consumption of a converted electric vehicle. SN Appl. Sci. **2**(4), 1–8 (2020)
19. Luo, Y., Tan, Y.P., Li, L.F.: Study on saving energy for electric auxiliary systems of electric bus. Energy Sources Part A Recovery Util. Environ. Eff. https://doi.org/10.1080/15567036.2020.1829750 (2020)
20. Li, P., Zhang, Y., Zhang, K., Jiang, M.: The effects of dynamic traffic conditions, route characteristics and environmental conditions on trip-based electricity consumption prediction of electric bus. Energy **218**, 119437 (2021)
21. Pamuła, T., Pamuła, W.: Estimation of the energy consumption of battery electric buses for public transport networks using real-world data and deep learning. Energies **13**(9), 2340 (2020)
22. Kivekäs, K., Vepsäläinen, J., Tammi, K.: Stochastic driving cycle synthesis for analyzing the energy consumption of a battery electric bus. IEEE Access **6**, 55586–55598 (2018)
23. Lajunen, A.: Energy consumption and cost-benefit analysis of hybrid and electric city buses. Transp. Res. Part C Emerg. Technol. **38**, 1–15 (2014)
24. Yi, Z., Smart, J., Shirk, M.: Energy impact evaluation for eco-routing and charging of autonomous electric vehicle fleet: ambient temperature consideration. Transp. Res. Part C Emerg. Technol. **89**, 344–363 (2018)
25. Iora, P., Tribioli, L.: Effect of ambient temperature on electric vehicles' energy consumption and range: model definition and sensitivity analysis based on nissan leaf data. World Electr. Veh. J. **10**(2), 1–15 (2019)
26. Liu, K., Wang, J., Yamamoto, T., Morikawa, T.: Exploring the interactive effects of ambient temperature and vehicle auxiliary loads on electric vehicle energy consumption. Appl. Energy **227**, 324–331 (2018)

Impacts Analysis of Rainfall on Road Traffic Operation

Shengyue Lyu, Wei Guo, and Yadan Yan

Abstract The rapid development of cities puts forward higher requirements for the resilience of the road network. As one of the unfavorable weathers, rainfall has a great impact on the operation of road traffic. Using weather and vehicle speed data of a city in China, this paper compares and analyzes changes in traffic operation indicators such as the standard speed and the percentage reduction of the standard speed under different rainfall intensities. Results indicate that under the same rainfall conditions and time period, the traffic operation of the elevated road is least affected by rainfall, and the traffic operation of underpass tunnels and flat roads is more affected by rainfall. With the increase of rainfall intensity, the reduction percentage of the standard speed of the underpass tunnel shows an increasing trend; however, it is less affected in light rain and moderate rain, and more affected in heavy rain and rainstorm. With the increase of rainfall intensity, the standard speed of the elevated road has no obvious change trend, and it is less affected by light rain and moderate rain, but more by rainstorm. In each time period, the standard speed reduction percentage of flat roads shows an obvious increasing trend with the increase of rainfall intensity.

Keywords Road traffic · Rainfall intensity · Standard speed · Road type

1 Introduction

According to the national natural disaster statistics for 2019 released by the Ministry of Emergency Management, waterlogging is one of the most destructive disasters faced by many cities [1]. As the lifeline system of the city, the road traffic system is an important part of the urban infrastructure. Under rain scenarios, road surface water may seriously affect the operation of urban road transportation systems, and even cause severe disruption of urban transportation networks [2]. With the advent

S. Lyu · Y. Yan (✉)
School of Civil Engineering, Zhengzhou University, Zhengzhou 450001, P. R. China
e-mail: yanyadan@zzu.edu.cn

W. Guo
Henan Transportation Research Institute Co., Ltd., Zhengzhou 450006, P. R. China

X. Qu et al. (eds.), *Smart Transportation Systems 2021*, Smart Innovation, Systems and Technologies 231, https://doi.org/10.1007/978-981-16-2324-0_11

of the era of big data, the acquisition of traffic data has brought more opportunities to the study of traffic problems. Continuous rainfall will cause great fluctuation in traffic operation. Zhou et al. proposed a recurrent neural network based microscopic car following model that is able to accurately capture and predict traffic oscillation [3]. Based on the V-K relationship of traffic flow, Pregnolato et al. developed a relationship between depth of standing water and vehicle speed [4]. Gong et al. analyzed the travel speed and traffic operation index under rainy weather and normal weather, carried out analysis from the perspectives of rainfall intensity, time period, congestion level, etc., and established a rainy weather speed prediction correction model [5]. Qu et al. identified that the speed-density relation could be well represented by single-regime models using a weighted least square method [6]. Then they applied a new calibration approach to generate stochastic traffic flow fundamental diagrams [7]. Su et al. studied the behavior of drivers in heavy rain, estimated the speed of vehicles under different water depth and visibility conditions, and then predicted the traffic conditions [8]. Kermanshah and Derrible studied the robustness of road networks to extreme flood events, using actual trips completed, changes in geographic information system attributes, and network topology indicators to characterize them [9]. Essien et al. discussed the impact of rainfall intensity on traffic flow parameters during peak and off-peak hours [10]. Ji and Shao analyzed the difference in free flow velocity between normal weather and heavy rain, and established a free flow velocity evolution model that varies with rainfall intensity under heavy rain using a fitting model [11]. Zhang and Alipour studied the degree of closure of roads and bridges under floods and real-time network characteristics to evaluate the user loss caused by increased traffic time [12]. In addition, Gao et al. examined the effects of irrational psychological inertia in mode shift behavior with controlling potential endogeneity [13]. Wu et al. proposed an emergency vehicles (EVs) lane pre-clearing strategy to prioritize EVs on such roads through cooperative driving with surrounding connected vehicles [14].

Most of existing studies have analyzed the impact of rainy weather on traffic flow parameters, and have not considered road types. Using the vehicle speed data, this paper studies the urban road traffic operation under rainfall scenarios, focusing on road segments with different vertical elevations. It is expected that this study could provide a reference for traffic management, as well as a reference for road users to formulate travel routes.

2   Data Preprocessing

The rainfall classification standard of the meteorological department is referred. The standard uses 24-h rainfall as the basis for dividing rainfall into four levels: light rain, moderate rain, heavy rain and rainstorm (Table 1). The vehicle speed data of 200 vehicles on 12 road segments in 12 days were used, covering four rainfall scenarios and three road types with different vertical elevations, as shown in Table 2. Three

Table 1 Date and weather condition in 2020

Weather	No rain	Light rain	Moderate rain	Heavy rain	Rainstorm
Data	Sep 25 Oct 7 Oct 14	Aug 14 Aug 15 Sep 21 Sep 27	Sep 23	Sep 19 Sep 26 Oct 13	Aug 12

Table 2 Road segments with three different types

Road type	Underpass tunnel	Plane road	Elevated road
Segments	Huakou Tunnel	Nanguo Middle Road	Foshan 1st Ring Highway
	Bigui Road Tunnel	Wenhua Road	Jiangwan Interchange
	Fenjiang Road Tunnel	South of Nanhai Avenue	Xingye Road Interchange
	Luocun Tunnel	Foshan Avenue	Luonan Bridge

time periods, i.e., morning peak period (7:00–9:00), off-peak period (12:00–14:00), and evening peak period (17:00–19:00) are selected for analysis.

In rainy weather, the road surface is slippery and the friction coefficient decreases. Drivers affected by the rainfall will have their sight blocked and their sight distance shortened, resulting in the inaccurate identification of the vehicle ahead and the signs and markings. In order to be safe, the driver often reduces the speed carefully. Especially when high-level rainfall occurs, road surface water will be produced, which will seriously affect the normal operation of the urban transportation system. Different road types are affected by precipitation differently. In particular, underpass tunnels are particularly serious due to their low elevation. On the contrary, high elevation segments such as elevated roads are not prone to water accumulation. Examples of different road types are shown in Figs. 1, 2 and 3.

3 Results Analysis

3.1 Indicators

The change of driving speed is an obvious indicator for the impact of rainfalls on traffic. In order to unify the evaluation standard, the average speed is standardized based on the speed limit value:

$$V = \frac{v}{v_{sl}} \tag{1}$$

Fig. 1 Huakou Tunnel (underpass tunnel)

Fig. 2 Nanguo Middle Road (plane road)

Fig. 3 Jiangwan Interchange (elevated road)

where V is the standard speed, and $0 \leq V \leq 1$; v is the average speed of vehicles, km/h; v_{sl} is the speed limit, km/h.

Table 3 Average speed of underpass tunnels

Road segment	Weather	v (km/h)		
		Morning peak	Off-peak	Evening peak
Huakou Tunnel	No rain	58.0	70.4	47.6
	Light rain	56.6	56.6	44.1
	Moderate rain	46.4	60.0	39.4
	Heavy rain	40.0	51.0	46.3
	Rainstorm	31.8	32.4	32.0
Bigui Road Tunnel	No rain	58.6	54.8	61.6
	Light rain	49.2	48.0	61.0
	Moderate rain	37.8	43.6	45.7
	Heavy rain	32.0	56.0	32.8
	Rainstorm	26.0	44.0	18.4
Fenjiang Road Tunnel	No rain	52.6	42.0	47.8
	Light rain	38.0	40.0	42.6
	Moderate rain	20.0	26.0	40.5
	Heavy rain	32.0	12.8	24.6
	Rainstorm	8.2	8.0	10.7
Luocun Tunnel	No rain	27.4	63.5	48.0
	Light rain	25.5	44.0	57.7
	Moderate rain	42.0	44.0	37.5
	Heavy rain	22.3	36.0	22.0
	Rainstorm	19.2	29.8	26.6

3.2 Calculation Results

As shown in Tables 3, 4 and 5, the average speed of three types of roads all shows an overall downward trend with the increase of rainfall intensity. But the average speed of some road segments increases during moderate rain and heavy rain. This may be because that heavy rainfall causes changes in people's travel demand, that is, fewer trips and lower road saturation. For further comparison, the average standard speed of different types of roads under different rainfall conditions is calculated (see Table 6).

The changes of standard speed under different time periods, road types and rainfall intensity are compared in Figs. 4, 5 and 6. Under various rainfall intensities, as a whole, the standard speed of the elevated road is least affected by rainfall at peak and off-peaks, while underpass tunnels and plane roads are more affected. Specifically, it can be found that:

(1) As the rainfall intensity increases, the percentage reduction of the standard speed of the underpass tunnels is increasing. However, it is less affected during

Table 4 Average speed of elevated roads

Road segment	Weather	v (km/h)		
		Morning peak	Off-peak	Evening peak
Foshan 1st Ring Highway	No rain	95.2	96.4	93.1
	Light rain	85.0	90.1	85.6
	Moderate rain	83.8	84.4	92.7
	Heavy rain	95.2	82.5	84.7
	Rainstorm	65.4	69.7	68.6
Jiangwan Interchange	No rain	37.5	40.0	22.6
	Light rain	28.6	39.1	18.7
	Moderate rain	26.5	38.6	19.4
	Heavy rain	26.4	26.0	15.9
	Rainstorm	20.0	23.8	12.8
Xingye Road Interchange	No rain	81.0	86.6	74.0
	Light rain	81.6	82.7	75.3
	Moderate rain	75.5	78.6	50.0
	Heavy rain	83.4	87.4	64.1
	Rainstorm	73.5	74.0	62.4
Luonan Bridge	No rain	84.0	80.1	81.3
	Light rain	82.1	77.0	78.3
	Moderate rain	86.0	80.0	73.7
	Heavy rain	88.3	73.0	80.6
	Rainstorm	79.9	69.3	71.4

light rain and moderate rain. During the morning peak, the reduction percentages of the standard speed are 6.8% and 6.7%, respectively. At the off-peak, the reduction percentages are 17.9% and 25.0%, respectively. At the evening peak, they are 2.1% and 18.2%, respectively. During heavy rain and rainstorm, the reduction percentage of the standard speed is greatly affected. The reduction percentages for the morning peak are 30.3% and 39.8%, respectively. At the off-peak, the reduction percentages are 34.7% and 53.2%, respectively. At the evening peak, the reduction percentages are 52.3% and 57.8%, respectively. The possible reason for the great reduction is the accumulation of water in the tunnel.

(2) The standard speed reduction percentage of the elevated road has no obvious trend with the increase of rainfall intensity. And it is less affected by light rain and moderate rain. The reduction percentages in the morning peak are 7.9% and 9.5%, respectively. At the off-peak, the reduction percentages are 4.5% and 6.4%, respectively. At the evening peak, the values are 5.4% and 12.8%, respectively. The reduction percentage of the elevated road is greatly affected by the heavy rain. The reduction percentage is 20.8% in the morning

Table 5 Average speed of plane roads

Road segment	Weather	v (km/h)		
		Morning peak	Off-peak	Evening peak
Nanguo Middle Road	No rain	54.9	46.8	40.7
	Light rain	47.8	40.0	37.6
	Moderate rain	38.9	40.0	40.0
	Heavy rain	36.1	36.0	19.8
	Rainstorm	15.0	20.0	14.0
Wenhua Road	No rain	18.4	27.0	16.0
	Light rain	15.9	28.3	20.0
	Moderate rain	17.6	20.0	13.6
	Heavy rain	11.7	12.0	11.7
	Rainstorm	10.8	13.4	7.8
South of Nanhai Avenue	No rain	43.6	28.3	25.7
	Light rain	38.5	20.0	26.3
	Moderate rain	34.9	22.1	24.2
	Heavy rain	36.7	18.6	13.6
	Rainstorm	32.5	17.7	14.3
Foshan Avenue	No rain	48.5	52.7	41.6
	Light rain	39.1	49.4	34.0
	Moderate rain	22.0	48.0	29.7
	Heavy rain	27.2	52.0	44.0
	Rainstorm	25.0	42.2	40.0

Table 6 Average standard speed of different types of roads

Road type	Time period	No rain	Light rain	Moderate rain	Heavy rain	Rainstorm
Underpass tunnel	Morning peak	0.514	0.479	0.479	0.368	0.309
	Off-peak	0.838	0.688	0.629	0.547	0.392
	Evening peak	0.549	0.537	0.450	0.314	0.232
Elevated road	Morning peak	0.859	0.792	0.778	0.833	0.680
	Off-peak	0.874	0.836	0.818	0.554	0.675
	Evening peak	0.766	0.725	0.688	0.690	0.604
Plane road	Morning peak	0.728	0.619	0.504	0.483	0.361
	Off-peak	0.701	0.633	0.584	0.519	0.241
	Evening peak	0.550	0.533	0.476	0.396	0.333

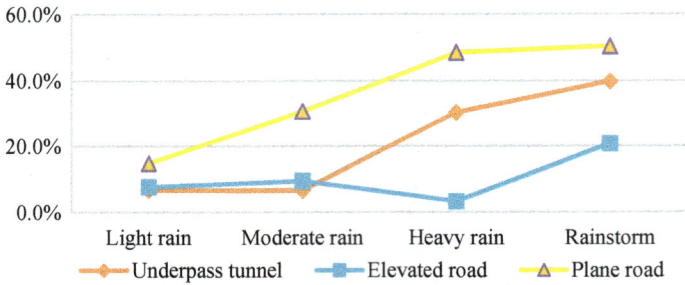

Fig. 4 Variation of standard speed reduction percentage (morning peak period)

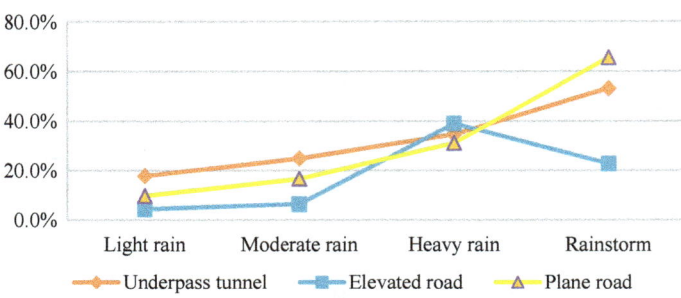

Fig. 5 Variation of standard speed reduction percentage (evening peak period)

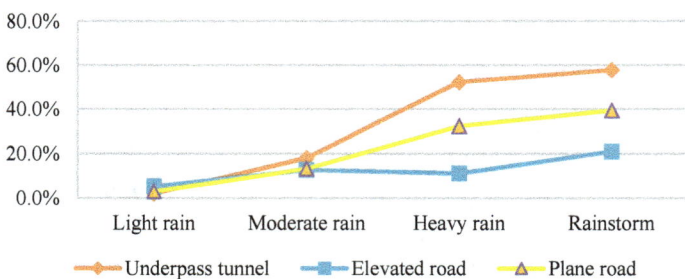

Fig. 6 Variation of standard speed reduction percentage (off-peak period)

peak, 22.8% in the off-peak, and 21.2% in the evening peak. The standard speed reduction percentage under heavy rain is greater than the speed reduction percentage under rainstorm, which may be due to the reduced travel demand.

(3) The standard speed reduction percentages of plane roads at various time periods shows an obvious trend of increasing with the increase of rainfall intensity. During the morning peak, the reduction percentage increases from 14.9% under the light rain condition to 50.4% under the rainstorm condition. At the off-peak, the value increases from 9.8% under the light rain condition to 65.7% under

the rainstorm condition, and the evening peak increased from 3.1% under the light rain condition to 39.4% under the rainstorm condition.

4 Conclusion

Based on the weather data and the driving speed data, the speed changes of three road types at different time periods and under different rainfall intensities are studied. However, there are still limitations for this work. Further studies can be carried out about the strategies from different perspectives to reduce the impacts of rainfall on the traffic operation.

Acknowledgements This work was supported by the National Natural Science Foundation of China (No. 51678535) and University Young Key Teacher Foundation by Henan Province (No. 2018GGJS004).

References

1. Fang, Y.: China's top 10 natural disasters in 2019. Life Disaster **252**(5), 14–15 (2020) (in Chinese)
2. Liu, W., Tang, Y., Yang, F., et al.: Research on urban transport network topology vulnerability identification under rainfall conditions. Int. J. Embedded Syst. **13**(3), 352–359 (2020)
3. Zhou, M., Qu, X., Li, X.: A recurrent neural network based microscopic car following model to predict traffic oscillation. Transp. Res. Part C Emerg. Technol. **84**, 245–264 (2017)
4. Pregnolato, M., Ford, A., Wilkinson, S., et al.: The impact of flooding on road transport: a depth-disruption function. Transp. Res. Part D Transp. Environ. **55**, 67–81 (2017)
5. Gong, D., Song, G., Li, M., et al.: Impact of rainfalls on travel speed on urban roads. J. Transp. Syst. Eng. Inf. Technol. **15**(1), 218–225 (2015)
6. Qu, X., Wang, S., Zhang, J.: On the fundamental diagram for freeway traffic: a novel calibration approach for single-regime models. Transp. Res. Part B Methodol. **73**, 91–102 (2015)
7. Qu, X., Zhang, J., Wang, S.: On the stochastic fundamental diagram for freeway traffic: model development, analytical properties, validation, and extensive applications. Transp. Res. Part B Methodol. **104**, 256–271 (2017)
8. Su, B., Huang, H., Li, Y.: Integrated simulation method for waterlogging and traffic congestion under urban rainstorms. Nat. Hazards **81**(1), 23–40 (2016)
9. Kermanshah, A., Derrible, S.: Robustness of road systems to extreme flooding: using elements of GIS, travel demand, and network science. Nat. Hazards **86**(1), 151–164 (2016)
10. Essien, A., Petrounias, I., Sampaio, P., et al.: The impact of rainfall and temperature on peak and off-peak urban traffic. In: Hartmann, S., Ma, H., Hameurlain, A., Pernul, G., Wagner, R. (eds.) Database and Expert Systems Applications. DEXA 2018. LNCS, vol. 11030, pp. 399–407. Springer, Regensburg (2018)
11. Ji, X., Shao, C.: Modeling and analyzing free-flow speed of environment traffic flow in rainstorm based on geomagnetic detector data. Ekoloji **28**(107), 4223–4230 (2019)

12. Zhang, N., Alipour, A.: Integrated framework for risk and resilience assessment of the road network under inland flooding. Transp. Res. Rec. **2673**(12), 182–190 (2019)
13. Gao, K., Yang, Y., Sun, L., et al.: Revealing psychological inertia in mode shift behavior and its quantitative influences on commuting trips. Transp. Res. Part F Traffic Psychol. Behav. **71**, 272–287 (2020)
14. Wu, J., Kulcsár, B.A., Ahn, S., et al.: Emergency vehicle lane pre-clearing: from microscopic cooperation to routing decision making. Transp. Res. Part B Methodol. **141**, 223–239 (2020)

Modeling Seafarer Change at Seaports in COVID-19

Yu Guo, Ran Yan, Yiwei Wu, and Shuaian Wang

Abstract Shipping is the most cost-effective way to transport large volumes of goods over long distances, and over 80% of goods traded worldwide are carried by sea. To keep our economy running, especially in difficult times of epidemics, such as the coronavirus disease 2019 (COVID-19), it is of utmost importance to keep ships sailing, moving the bulk of goods including medical supplies and food. Every month, around 100,000 seafarers need to disembark from the ships that they operate to comply with regulations governing safe working hours and crew welfare, and another 100,000 seafarers will embark to continue to move the global trade safely. However, as countries around the world tighten border controls in against the spread of the coronavirus outbreak, seafarers are prohibited from boarding or leaving ships at most ports, with the exception of just a few. This situation is leading seafarers to serve onboard vessels beyond their contracted shifts. Given that more than a quarter of seafarers suffer from depression because of their long time spent at sea and being away from family and friends, banning crew changes will put their mental health at risk. This will further increase the likelihood of marine accidents, jeopardize global supply chains, and ultimately exacerbate current hardships. To tackle this emergency, the International Maritime Organization and the European Commission, amongst others, call on governments to coordinate efforts to designate ports for crew changes while preventing the potential spread of the coronavirus. The aim of the study is to develop a framework that integrates big data, machine learning, and operational research for governments and supranational organizations to choose ports for crew changes, safeguarding seafarers' welfare, and strengthening the global response to the threats of epidemics.

Keywords Shipping · Seafarer change · Epidemics · At seaports · World trade · Framework

Y. Guo (✉) · R. Yan · Y. Wu · S. Wang
Department of Logistics and Maritime Studies, Hong Kong Polytechnic University, Kowloon, Hong Kong
e-mail: christineyu.guo@connect.polyu.hk

© The Author(s), under exclusive license to Springer Nature Singapore Pte Ltd. 2021   115
X. Qu et al. (eds.), *Smart Transportation Systems 2021*, Smart Innovation,
Systems and Technologies 231, https://doi.org/10.1007/978-981-16-2324-0_12

1 Introduction

Shipping is the backbone of international trade. The United Nations estimated the annual global volume of seaborne shipments in 2018 at 11 billion tons [1]. In the European Union (EU), 75% of the international trade of goods and 30% of goods circulating in the internal market are transported by sea [2]. For island countries such as the United Kingdom (UK), Japan, and New Zealand, almost 100% of exports and imports are fulfilled by ships. In the time of a global crisis, such as the COVID-19, it is more important than ever to allow maritime trade moving to sustain global supply chains.

The world's merchant fleet at sea is operated by two million seafarers. Seafaring contracts typically last between three and nine months, followed by unpaid time ashore. Crew members work for long hours in a day and seven days a week. Stress, anxiety, and depression are common among those who spend a long time at sea, away from family and friends. Research from Yale University in 2018 revealed that more than a quarter of seafarers suffered from depression [3], while a survey last year conducted by Yale University showed 20% had either considered suicide or attempted suicide [4].

Crew changes are essential to functioning international shipping industry. About 100,000 seafarers need to change over on a monthly basis to comply with maritime and labor regulations, according to estimates of the International Chamber of Shipping and the International Transport Workers' Federation. However, with most of the world in lockdown mode in a bid to limit the spread of coronavirus, seafarers are stuck on board because they are denied the right to board or leave ships at many ports. This situation is leading seafarers to serve onboard vessels beyond their contracted shifts, putting their physical and mental health at risk. Considering that a quarter of seafarers already suffer from depression, prolonging their duration of work at sea will exponentially raise the probability of marine accidents, a disaster for an already fragile and stretched global economy.

A handful of authorities have taken efforts to alleviate the problem. The UK is by far the only country that categorized seafaring as an essential service. Ports in Canada have remained open for crew changes. China and Singapore have started to allow crew changes on a case-by-case basis. Nevertheless, these fragmented efforts are insufficient to ensure the movement of seafarers around the world.

To tackle this emergency, number of industrial associations and intergovernmental organizations have called on governments to coordinate efforts to designate ports for crew changes while preventing the potential spread of the coronavirus. The International Maritime Organization (IMO), a body of the United Nations, has called for governments to put in place pragmatic and coordinated arrangements to allow crew changes to take place. The International Chamber of Shipping and International Transport Workers' Federation has stressed that the shipping industry urgently needs governments and administrations to coordinate their efforts to facilitate the changeover of ships' crew. In April 2020, the European Commission (EC) called on its Member States to coordinate efforts to designate several ports in the EU for crew

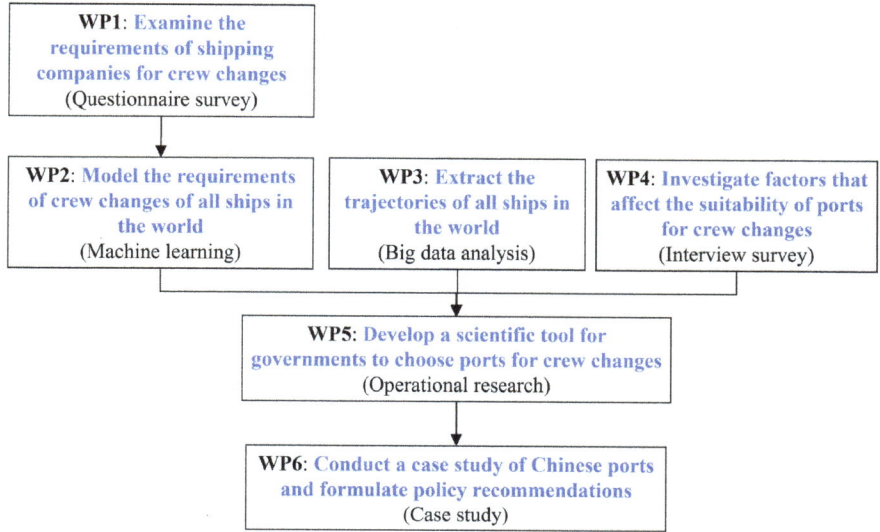

Fig. 1 Framework of the study

changes. China's Ministry of Transport has issued guidance on coordinating local governments and port authorities to allow crew change yet targeting only Chinese seafarers working on international-trading ships to date.

The aim of the study is to develop a framework of a scientific tool that integrates big data, machine learning, and operational research for governments and supranational organizations to choose ports for crew changes, and thus to safeguard seafarers' welfare and strengthen the global response to the threats of epidemics. The overall framework of this work is presented in Fig. 1.

2 Literature Review

The existing research focuses on the physical and mental stress of seafarers and on the shortage of seafarer supplies. Researchers from Cardiff University [5] surveyed 1856 seafarers and found that seafarers suffered from excessive working hours, often more than 12 h a day, hidden by a concerning number of falsified audited records; furthermore, over 40% of the seafarers reported disturbed sleep, mainly due to noise and motion. An Australian team [6] summarized maritime accident data from 1960 to 2009 and pointed out that over 9000 seafarers committed suicide or disappeared at sea, which proves that the mental health of seafarers is poor and often fatal. This result is echoed by research conducted by Yale University in 2019, which showed 20% of seafarers had either considered suicide or attempted suicide [4]. Because of the chronic physical and mental fatigue, overwork time, and isolation from families and friends at sea, seafarers are moving away from maritime jobs and it is difficult for the industry to attract sufficient new laborers. The International Chamber of Shipping, based on their maritime manpower report, forecasted that there will be a shortage of 147,500 senior seafarers by 2025 [7].

The global large-scale denial of crew changes occurring in 2020 poses an unprecedented challenge since World War II. There is a pressing need to develop scientific and innovative approaches for tackling the problem by choosing ports in a holistic and coordinated manner for crew changes. This study will adopt a framework consisting of multidisciplinary approach by taking advantage of big data, machine learning, and operational research to address this problem.

3 Work Packages and Planned Tasks (Gantt Chart) and Methodology

This framework consists of six work packages (WP), as shown in Fig. 1. The details of each work package are presented as follows.

WP1: Examine the requirements of shipping companies for crew changes (Questionnaire survey)

A questionnaire survey will be conducted among shipping companies to solicit their actual crew change requirements.

Task 1.1. Questionnaire design. The questionnaire survey will be designed to solicit (i) ships' features (type, tonnage, age, flag states, etc.), and (ii) seafarers' features (number, nationality, contract duration, preferred ports for embarkation and disembarkation). The designed questionnaire will first be circulated to a small group of respondents to obtain their feedback and then a revised questionnaire will be sent out.

Task 1.2. Administer survey. The questionnaire survey will be administered to collect shipping companies' responses covering no fewer than 100 ships. Marine government agencies will be contacted to help promote the survey. The respondents will be contacted via telephones, emails, and WhatsApp. Online survey tools will be used to facilitate data collection.

WP2: Model the requirements of crew changes of all ships in the world (Machine learning)

The requirements for crew changes obtained by the survey in WP1 is only a small sample of all ships in the world. It is impossible to obtain crew change requirements data for all ships of the world. We will, therefore, utilize the survey data obtained in WP1 and develop a random forest based supervised machine learning model to predict the crew change requirements of ships based on ships' features such as ship type, tonnage, age, and flag states. A database of type, tonnage, age, flag states of the world's ships is available in the library of the principal investigator's institute. We will then use the machine learning model to predict the crew change requirements of all ships in the world.

WP3: Extract the trajectories of all ships in the world (Big data analysis)

Seafarers can embark or disembark only when their ship is calling at a port. The crew change requirement of a ship in WP1 and WP2, for example, seafarers from Indonesia prefer embarking or disembarking at a port in Southeast Asia, must be combined with the actual trajectories of ships (including when and which ports are visited) to serve as inputs for governments to designate ports for crew changes. The automatic identification system (AIS) is an automatic tracking system compulsory for all ships of 300 gross tonnage and upwards, recording the locations of ships at different timestamps which are transmitted to other ships, ports, and maritime authorities. The volume of one year's AIS data is several terabytes (TB). We will apply big data analysis techniques to merge the AIS big data with port data (name, country, location) to extract one year's trajectory of each ship in the world's fleet.

WP4: Investigate factors that affect the suitability of ports for crew changes in face of epidemics (Interview survey)

Ports that are suitable for crew changes in situations of epidemics should have accommodation areas for seafarers to quarantine and should be connected to operational airports and rail stations to allow for swift travel and repatriations of seafarers. To obtain a holistic view of the current practices for choosing a port for crew changes, we will conduct 20 face-to-face structured interviews with target interviewees drawn from ports, customs, maritime authorities, and transport authorities. Due to limited inspection and quarantine resources, a government can only open a limited number of ports for crew changes. We will develop a tool for governments to choose the ports for crew changes in a coordinated manner.

WP5: Develop a scientific tool for governments to choose ports for crew changes (Operational research)

The outputs of WP2, WP3, and WP4 will be the input of a scientific tool for governments to identify ports where crew changes can be organized. For example, if China plans to open three ports for crew changes, which three ports should be chosen to maximize the total number of seafarers changed? Intuitively, major ports with a large number of ship calls should be prioritized. However, in reality there are many factors that affect whether a port should be chosen. For instance, Ningbo-Zhoushan is the largest port in the world in terms of cargo throughput, however, there is no major airport nearby and it is inconvenient for seafarers to disembark at Ningbo-Zhoushan and go home. In general, the chosen ports should be geographically dispersed so as to cover a large hinterland. Nevertheless, there are too many ports to choose from in reality. For instance, the UK has 260 ports and China has 157 ports. It is practically impossible to enumerate all possible combinations of port choices. To address this challenge, we will use operational research to develop mathematical models to generate optimal decisions for authorities.

WP6: Conduct a case study of Chinese ports and formulate policy recommendations (Case study)

China is the world's largest supplier of seafarers. China owns the third-largest tonnage of ships, and seven of the world's top ten ports in terms of cargo throughput are located in China. Therefore, crew changes are of vital significance to China. Therefore, a case study of Chinese ports will be conducted, and the same method can be applied to other governments and supranational organizations.

Task 6.1. Case study. A case study based on Chinese ports and all ships in the world that visit Chinese ports will be conducted to validate the effectiveness of the developed port choice tool.

Task 6.2. Focus group meeting. We will invite representatives of maritime authorities, shipping companies, and seafarers to take part in a focus group meeting to present the study findings to stakeholders and solicit their comments. Policy recommendations will be formulated for maritime authorities.

We will summarize research findings, including guidelines for maritime authorities, the tool for choosing ports for crew changes, and the results of case study. An online platform will be developed to share the research findings so that governments and shipping professionals can freely download the tool and tailor it to their needs. We will present our findings to the Ministry of Transport of China and invite the International Maritime Organization to promote the research findings among maritime authorities.

4 Impact of the Work

4.1 Impact on Insurance Business/Main Risk

Seafarers are vulnerable to epidemics. Once they contract an epidemic at sea, it is difficult for them to receive timely medical treatment. Due to the international nature of shipping, most seafarers have to disembark at a port in a foreign country. However, in case of an epidemic, many countries will shut down their borders, leaving the seafarers stranded at sea for periods far beyond their contracts, no knowing when they can return home. Considering that in normal situations seafarers already have to work in an isolated environment for several months and suffer from stress, anxiety and depression, prohibiting them from leaving a ship will be detrimental to their mental health. A scientific tool for governments and supranational organizations to choose ports for crew changes will contribute to protecting the welfare of 1.6 million seafarers from epidemics.

4.2 Impact on Society

Shipping the most cost-effective mode of transporting large volumes of goods over long distances. Over 80% of goods traded worldwide are carried by sea, including food, medical supplies, energy, raw materials and manufactured products. In epidemics, a number of production activities are stopped to prevent the transmission of diseases. Then, shipping plays a more vital role to sustain global supply chains and provide many countries with essential food and medical supplies. Facilitating crew changes will enable them to continue keeping world trade flowing, strengthening the global response to epidemics.

5 Conclusion

As the economic development and world trade depend largely on maritime transportation, it is important to continue transporting cargos in COVID-19. The world's merchant fleet at sea is operated by two million seafarers. Seafaring contracts typically last between three and nine months, followed by unpaid time ashore. Working long hours every day in ships, crew members fell anxiety, stress and depression causing 20% has potential risk to suicide. Hence, it is essential to crew changes over on a monthly basis to comply with maritime and labor regulations. However, seafarers are stuck on board with most of the world in lockdown mode in a bid to limit the spread of coronavirus which means it is difficult to crew changes. In order to tackle this emergence, IMO has called for governments to allow crew changes to take place.

In this work, we aim to develop a framework of a scientific tool that integrates big data, machine learning, and operational research for governments and supranational organizations to choose ports for crew changes, and thus to safeguard seafarers' welfare and strengthen the global response to the threats of epidemics. In the framework WP1, we collect information through questionnaire which include ships' features and seafarers' features to examine crew changes requirements. Because data is small in WP1. Hence, we utilize the survey data obtained in WP1 and develop a random forest based supervised machine learning model to predict the crew change requirements in WP2. We need to extract trajectories of all ships. Because the crew change requirement of a ship in WP1 and WP2 must be combined with the actual trajectories of ships. Due to limited inspection and COVID-19, we will develop a tool to choose the ports for crew changes in a coordinated manner. How to choose ports is a problem. To address this challenge, we will use operational research to develop mathematical models to generate optimal decisions. Finally, we will conduct a case study because crew changes are of vital significance to China.

Crew changes will have good impact on seafarers and society. From the seafarers' perspective of a scientific tool to choose ports for crew changes could contribute to protecting the welfare of 1.6 million seafarers from epidemics. Form the perspective of society, facilitating crew changes will enable them to continue keeping world trade flowing, strengthening the global response to epidemics. Because shipping plays a more vital role to sustain global supply chains.

References

1. United Nations Conference on Trade and Development: Review maritime transportation. Paper presented at the United Nations Conference on Trade and Development, New York and Geneva (2019)
2. European Commission: Commission presents guidelines on health, travel arrangements and repatriation of persons on board ships. https://ec.europa.eu/commission/presscorner/detail/en/QANDA_20_632 (2020). Accessed 20 Apr 2020
3. Safety 4 Sea: 26% of seafarers show signs of depression. https://safety4sea.com (2018). Accessed 23 Apr 2020
4. Lefkowitz, R.Y., Slade, M.D.: Seafarer mental health study. https://www.seafarerstrust.org/wpcontent/uploads/2019/11/ST_MentalHealthReport_Final_Digital-1.pdf (2019). Accessed 23 Apr 2020
5. Smith, A.P., Allen, P.H., Wadsworth, E.J.K.: Seafarer fatigue. https://www.seafarerstrust.org/wpcontent/uploads/2019/11/ST_MentalHealthReport_Final_Digital-1.pdf (2006). Accessed 25 Apr 2020
6. Iversen, R.T.: The mental health of seafarers. Int. Marit. Health **63**(2), 78–89 (2012)
7. Safety4Sea: Global supply and demand for seafarers. https://safety4sea.com (2018). Accessed 27 Apr 2020

Driving Style Recognition Incorporating Risk Surrogate by Support Vector Machine

Qingwen Xue, Jianjohn Lu, and Kun Gao

Abstract Accurate driving style recognition is a crucial component for advanced driver assistance systems and vehicle control systems to reduce potential rear-end collision risk. This study aims to develop a driving style recognition method incorporating matching learning algorithms and vehicle trajectory data. A risk surrogate, Modified Margin to Collision (MMTC), is proposed to evaluate the collision risk level of each driver's trajectory. Particularly, the traffic level is considered when labelling the driving style, while it has a great impact on driving preference. Afterwards, each driver's driving style is labelled based on their collision risk level using the K-means algorithm. Driving behavior features, including acceleration, relative speed, and relative distance, are extracted from vehicle trajectory and processed by time-sequence analysis. Finally, Supporting Vector Machine (SVM) is applied to recognize driving style based on the extracted features and labelled data. The performance of Random Forest (RF), K-Nearest Neighbor (KNN), and Multi-Layer Perceptron (MLP) are also compared with SVM. The "leave-one-out" method is used to validate the performance and effectiveness of the proposed model. The results show that SVM over performs others with 91.7% accuracy. This recognition model could effectively recognize the aggressive driving style, which can better support ADAS.

Keywords Driving style recognition · Vehicle trajectory · Risk surrogate · Key feature extraction · SVM

1 Introduction

Driving style refers to how drivers choose to drive habitually and the driver states that represent the common parts of varied driving behavior [1]. Recognition of a driver's

Q. Xue · J. Lu
The Key Laboratory of Road and Traffic Engineering, Ministry of Education, Tongji University, Shanghai, China

Q. Xue · K. Gao (✉)
Department of Architecture and Civil Engineering, Chalmers University of Technology, Goteborg, Sweden
e-mail: gkun@chalmers.se

X. Qu et al. (eds.), *Smart Transportation Systems 2021*, Smart Innovation, Systems and Technologies 231, https://doi.org/10.1007/978-981-16-2324-0_13

driving style based on rear-end collision risk is of great significance to improve driving safety. It is important to guarantee the safety and adequate performance of drivers and essential to meet drivers' needs, adjust to the drivers' preference, and ultimately improve the driving environment's safety. Driving style recognition also has potential value to help agencies effectively design control strategies [2, 3].

The paper proposed a driving style recognition model to consider the impact of traffic flow levels on driving behavior. The traffic flow level could be classified into normal and congested traffic, and the driving style is labelled concerning the road condition. The difference of risk surrogates between normal and congested traffic would be explored. The decision about driving style ignoring the traffic level is not acceptable. Therefore, the traffic flow levels would be taken into consideration when labelling and recognizing the driving style. The trajectory data extracted from the video are studied in the paper, which contains the identification, GPS, velocity, and acceleration.

2   Literature Review

In recent years, the studies about new modes of travel transportation [4, 5] and innovative approaches have been developed. Some research concentrates on the cooperative schedule to achieve the optimization [6, 7]. Machine learning algorithms applied to driving behavior recognition have been studied in some previous works. Different types of neural network (NN) algorithms have been used [8–11]. However, a larger size of the network could lead to a long training time [12]. The tree-like algorithm [13, 14] and Hidden Markov Model (HMM) [15, 16] are also adopted to detect the driving behaviors according to the extracted features. Some researchers also combined the HMM with dynamic Bayesian networks or ANN to predict the driving behavior by learning the driving data [17, 18]. While HMM requires a long training time, especially for a high number of states, the recognition time also increases with the number of states [19]. Therefore, a more suitable and effective method should be found to identify the driving style. SVM has been widely applied to various kinds of pattern recognition problems, including voice identification, text categorization, face detection [20, 21]. In addition, SVM performs well with a limited number of training samples, and SVM has fewer parameters to be determined [22, 23]. Therefore, many studies employed SVM to build driving style recognition models.

Except for unsupervised machine learning algorithms, for example, clustering, other machine learning algorithms require labelled or partially labelled driving behavior data. Some research adopted behavior-based or accident-based method to label the driving style [20, 24]. Driver self-reported questionnaire [25] and expert scoring [13] are also adopted to evaluate driving style. However, these two methods rely on drivers or experts' subjective judgments and can be very time-consuming when the number of drivers in the sample is huge. This paper proposes a new driving data label method based on collision surrogates incorporating traffic level.

3 Methodology

Three collision risk surrogates are used to determine the risk level of the car-following process for each following pairs. The threshold value to classify the risk level is different for normal and congested traffic. The K-means algorithm is applied to group the drivers as calm or aggressive, based on their trajectory risk levels. As the traffic flow has a great impact on driving behavior, it is considered when labelling the driving style.

3.1 Collision Risk Surrogate

It is essential to find the most effective surrogates to describe the collision risk when driving on the road [26–29]. Vehicle trajectory data such as the vehicle's velocity and acceleration are usually not good enough to estimate the rear-end collision risk. In the paper, the Margin to Collision (MTC) is used to evaluate the risk.

MTC indicates the final relative position of PV and FV if two vehicles decelerate abruptly.

$$MTC = (x_r + v_p^2/2a_p)/(v_f^2/2a_f) \tag{1}$$

where a_f and a_p denote the deceleration for FV and PV, respectively. Usually, both are defined as 0.7G. v_f and v_p respectively denote the velocity of FV and PV. x_r denotes relative distance. A modified MTC (MMTC) is proposed in the paper to include the following vehicle's reaction time when the PV abruptly decelerates. The equation is modified as follows.

$$MMTC = (x_r + v_f^2/2a_f - v_p^2/2a_p)/v_f \tag{2}$$

MMTC evaluates the minimum reaction time needed for FV to avoid a collision when PV abruptly decelerates. The collision risk is higher with a lower MMTC value since drivers have little time to react. MMTC can evaluate potential collision risk with the abrupt deceleration of PV.

3.2 Key Features Extraction

In this paper, the vehicle acceleration a_f, relative distance x_r, and relative velocity v_r are adopted to recognize the driving style. The Discrete Fourier Transform (DFT), and Statistical method (SM) are used respectively to extract the effective key features from the vehicle trajectory. The key parameters that can capture most of the distribution information of vehicle trajectory.

3.3 Recognition Algorithms

Four machine learning algorithms, i.e., SVM, RF, KNN, and MLP, are adopted to build the driving style recognition model. The inputs of the model are the features extracted in Sect. 3.2. The output of the model is the driving style. The recognition process of driver's aggressive driving style is as follows:

Step 1: Use "leave-one-out" to divide the test set and training set for the model. Select one sample as the training set and the others are test set, ensuring the training set contains calm and aggressive driving styles.

Step 2: In order to avoid the influence of dimension among different trajectory variables and eliminate the differences, the min–max normalization method is used to normalize the sample data.

Step 3: The Differential Evolution algorithm (DE) is applied to optimize the parameters of algorithms, and get the initial structure value of the optimized algorithm.

Step 4: Four algorithms are used to identify the aggressive driving style under normal and aggressive traffics.

Step 5: Model performance evaluation.

4 Results and Discussion

In this paper, the I-80 trajectory dataset of Next Generation Simulation (NGSIM) is adopted to study driving style. According to the data analysis, the aggregate flows of HOV lanes are respectively 250 and 398 vph during two periods, indicating two levels of traffic flow. 370 Leader-follower Vehicle Pairs (LVP) are chosen under congested and normal traffic to study the driving style in this paper since fewer interrupting vehicles are from other lanes.

4.1 Significant Analysis Considering Traffic Levels

Based on the trajectory data extracted from NGSIM, the significant analysis between trajectory features considering traffic levels has been conducted, shown in Table 1. Table 1 shows that there is no significant difference of ITTC surrogate in different traffic levels. Besides that, the drivers tend to keep higher velocity (17.314 m/s) and lower acceleration (0.040 m/s^2) when following preceding vehicles. And drivers in congested traffic flow keep higher velocity difference with preceding vehicles to keep safe, while the gap is smaller. However, the ITTC, THW, and MMTC are smaller

Table 1 Significant analysis considering traffic level

Variables	Description	Normal		Congested		P
		Mean	SD	Mean	SD	
v_f	The velocity of following vehicle (m/s)	17.31	3.76	13.80	3.20	0.000**
a_f	The acceleration of following vehicle (m/s^2)	0.04	1.93	0.06	1.69	0.005**
v_r	The difference of velocity between two following vehicles (m/s)	0.02	2.52	0.138	2.33	0.000**
x_r	The gap between two following vehicles (m)	37.75	28.97	25.90	16.07	0.000**
THW	Time to headway (s)	2.20	1.79	1.88	1.08	0.000**
ITTC	Inversed time to collision (s^{-1})	−0.01	2.57	−0.06	15.95	0.373
MTC	Margin to collision (s)	2.24	2.08	1.94	1.46	0.000**

**Significant correlation at 0.01 level (bilateral)

for congested traffic condition. Therefore, the traffic condition should be taken as a reference when labelling the risky following maneuvers.

4.2 The Sample Data Label

Figure 1 shows the fitting curves of MMTC by adopting three distributions for normal and congested traffic flow, i.e., normal distribution, logistic distribution, and t distribution. The t distribution achieves a better fitting performance than the other two distributions. Therefore, the t distribution is adopted to determine the threshold value of features. The 85% percentile value based on cumulative distribution are also obtained as threshold to classify the car-following maneuvers into several segments with two levels of risk, i.e., safe and risky.

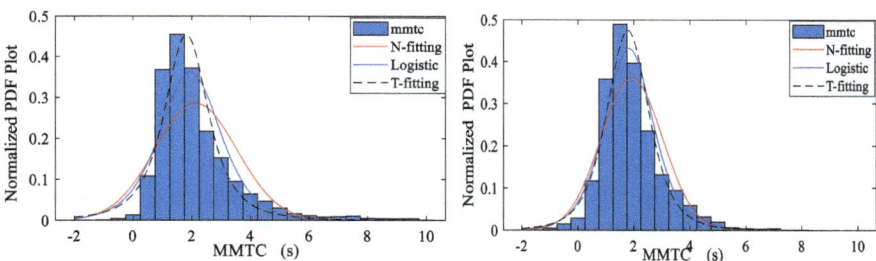

Fig. 1 Statistical fitting curves for MMTC for normal traffic flow and congested traffic flow

Fig. 2 Trajectory segments for a driver based on threshold values of MMTC

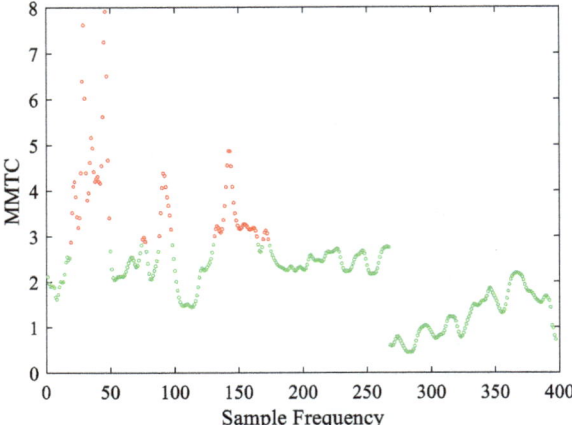

Each driver's driving trajectory can be divided into several segments, which belong to two different driving risk levels. A driver is selected to show the trajectory segments according to the threshold values of MMTC, shown in Fig. 2.

The proportions of trajectory segments with different risk levels can determine each driver's driving style. The K-means algorithm is applied to trajectory under different traffic flow to classify the drivers as calm and aggressive based on the ratio of risky maneuver. The clustering results show there are 246 calm drivers and 124 aggressive drivers under normal traffic flow, and 200 calm drivers and 170 aggressive drivers under congested traffic flow.

4.3   Driving Style Recognition

The SVM method is adopted to recognize the driving style under each traffic level. In this paper, the trajectory data including the vehicle acceleration a_f, relative distance x_r, and relative velocity v_r are adopted to recognize the driving style, respectively. The Discrete Fourier Transform (DFT) and Statistical method (SM) are respectively used to extract the effective key features from the vehicle trajectory. The z-score method is adopted to standardize features before model training.

In the study, the accuracy, precision, and recall rates are assessed to evaluate the model's ability to recognize aggressive drivers among all vehicles. The performance of the recognition model is evaluated using the "leave-one-out" cross-validation method. Driving style recognition results based on different feature extraction methods and SVM are shown in Table 2. Except mentioned, the SVM algorithm uses the linear kernel function.

As shown in Table 2, the DFT outperforms SM in feature extraction, with an accuracy of 91.7%. With any combinations of features, the accuracy rate of the SM is lower than that based on DFT. In general, the feature x_r and v_r perform better than

Table 2 The recognition results of driving style based on SVM with DTF and SM

Features	DFT (%)	SM (%)
af	67.0	66.2
vr	83.2	79.7
xr	88.9	83.5
af + vr	83.2	78.7
vr + xr	87.8	83.2
af + xr	88.9	84.6
af + xr + vr	91.7	85.7

Table 3 The recognition results of driving style based on RF

Features	Algorithm	Accuracy (%)	Precision (%)	Recall (%)
af + xr + vr	RF	91.6	89.2	82.0
	MLP	88.1	73.0	87.3
	KNN	87.6	85.7	76.2
	SVM	91.7	92.8	81.8

Table 4 The recognition results of driving style with and without traffic level

Features		Accuracy (%)	Precision (%)	Recall (%)
af + xr + vr	With context	91.7	92.8	81.8
	No context	76.50	87.5	68.4

a_f in recognizing the driving style. A possible reason is that the rear-end collision risk determines the driving style label, the feature a_f cannot accurately describe the relative motivation between two following vehicles.

The performance of four machine learning algorithms RF, MLP, KNN and SVM using all features with DFT method are compared. The accuracy, precision, and recall rates are listed in Table 3. It can be seen that SVM outperforms other machine learning algorithms. Random Forest is the second best algorithm. However, MLP gives the highest recall rate among all candidates. KNN, as the simplest classification method, unsurprisingly obtains the worst performance. As seen from Table 4, the recognition accuracy of driving style without traffic levels is 76.50%, lower than that with context. Therefore, the traffic levels should be taken into consideration.

5 Conclusion

In this study, a novel driving style label method is proposed to assign calm and aggressive labels based on collision risk incorporating traffic levels, which is critical to sample data needed in supervised machine learning. The main findings could be summarized as follows:

The rear-end collision risk surrogates, namely MMTC, are adopted to evaluate the risk during the car-following process. Each driver's trajectory can be divided into two risk levels incorporating traffic levels, and all drivers can be grouped into calm or aggressive using the K-means algorithm.

Two feature extraction methods are compared in the recognition model. Three machine learning algorithms including RF, MLP, and KNN are also adopted to compare with the SVM. The results show that DFT could better capture the characteristics of driving behavior. The driving style can be recognized with the highest accuracy of 91.7% using SVM.

This study offers the possibility of developing more sophisticated driving style recognition methods. For further work, the proposed method can be extended by selecting other features that can reflect the driving style more accurately. As we know, the driving style is also influenced by the road conditions. Such results can also be used to improve driving style recognition.

References

1. Gao, K., Tu, H., Sun, L., et al.: Impacts of reduced visibility under hazy weather condition on collision risk and car-following behavior: implications for traffic control and management. Int. J. Sustain. Transp. **14**(8), 635–642 (2020)
2. Zhou, M., Qu, X., Li, X.: A recurrent neural network based microscopic car following model to predict traffic oscillation. Transp. Res. Part C Emerg. Technol. **84**, 245–264 (2017)
3. Zheng, Y., Li, S.E., Wang, J., et al.: Stability and scalability of homogeneous vehicular platoon: study on the influence of information flow topologies. IEEE Trans. Intell. Transp. Syst. **17**(1), 14–26 (2015)
4. Gao, K., Yang, Y., Li, A., Li, J., Yu, B.: Quantifying economic benefits from free-floating bike-sharing systems: a trip-level inference approach and city-scale analysis. Transp. Res. Part A Policy Pract. **144**, 89–103 (2021)
5. Xu, Y., Zheng, Y., Yang, Y.: On the movement simulations of electric vehicles: a behavioral model-based approach. Appl. Energy **283**, 116356 (2021)
6. Gao, K., Yang, Y., Sun, L., Qu, X.: Revealing psychological inertia in mode shift behavior and its quantitative influences on commuting trips. Transp. Res. Part F Traffic Psychol. Behav. **71**, 272–287 (2020)
7. Wu, J., Kulcsár, B., Ahn, S., et al.: Emergency vehicle lane pre-clearing: from microscopic cooperation to routing decision making. Transp. Res. Part B Methodol. **141**, 223–239 (2020)
8. Molchanov, P., Gupta, S., Kim, K., et al.: Multi-sensor system for driver's hand-gesture recognition. In: IEEE International Conference and Workshops on Automatic Face and Gesture Recognition, pp. 1–8. IEEE (2015)
9. Hashemi, A., Saba, V., Resalat, S.N.: Real time driver's drowsiness detection by processing the EEG signals stimulated with external flickering light. Basic Clin. Neurosci. **5**(1), 22 (2014)
10. Wang, H., Zhang, C., Shi, T., et al.: Real-time EEG-based detection of fatigue driving danger for accident prediction. Int. J. Neural Syst. **25**(02) (2015)
11. Srinivasan, D., Jin, X., Cheu, R.L.: Adaptive neural network models for automatic incident detection on freeways. Neurocomputing **64**(1), 473–496 (2005)

12. Singh, N.K., Singh, A.K., Tripathy, M.: A comparative study of BPNN, RBFNN and ELMAN neural network for short-term electric load forecasting: a case study of Delhi region. In: International Conference on Industrial and Information Systems. IEEE (2015)
13. Bejani, M.M., Ghatee, M.: A context aware system for driving style evaluation by an ensemble learning on smartphone sensors data. Transp. Res. Part C Emerg. Technol. **89**, 303–320 (2018)
14. Li, G., Li, S.E., Cheng, B., et al.: Estimation of driving style in naturalistic highway traffic using maneuver transition probabilities. Transp. Res. Part C Emerg. Technol. **74**, 113–125 (2017)
15. Berndt, H., Dietmayer, K.: Driver intention inference with vehicle onboard sensors. In: 2009 IEEE International Conference on Vehicular Electronics and Safety (ICVES), pp. 102–107. IEEE (2009)
16. Meng, X., Lee, K.K., Xu, Y.: Human driving behavior recognition based on hidden Markov models. In: Proceedings of the IEEE International Conference on Robotics and Biomimetics (2007)
17. Kumagai, T., Sakaguchi, Y., Okuwa, M., et al.: Prediction of driving behavior through probabilistic inference. In: Proceedings of the Eighth International Conference on Engineering Applications of Neural Networks (2003)
18. Zong, C.F., Yang, X., Wang, C., et al.: Driver's driving intention identification and behavior prediction during vehicle steering. J. Jilin Univ. (Eng. Technol. Ed.) **s1**, 27–32 (2009)
19. Aoude, G.S., Desaraju, V.R., Stephens, L.H., et al.: Driver behavior classification at intersections and validation on large naturalistic data set. IEEE Trans. Intell. Transp. Syst. **13**(2), 724–736 (2012)
20. Chen, Z., Wu, C., Huang, Z., et al.: Dangerous driving behavior detection using video-extracted vehicle trajectory histograms. J. Intell. Transp. Syst. **21**(11) (2017)
21. Sun, W., Zhang, X., Peeta, S., et al.: A real-time fatigue driving recognition method incorporating contextual features and two fusion levels. IEEE Trans. Intell. Transp. Syst. **18**(12), 3408–3420 (2017)
22. Feng, L., Yao, Y., Jin, B.: Research on credit scoring model with SVM for network management. J. Comput. Inf. Syst. **6**(11), 1032–1040 (2010)
23. Martinez, C.M., Heucke, M., Wang, F.Y., et al.: Driving style recognition for intelligent vehicle control and advanced driver assistance: a survey. IEEE Trans. Intell. Transp. Syst. **19**(3), 666–676 (2017)
24. Teimouri, F., Ghatee, M.: A real-time warning system for rear-end collision based on random forest classifier. arXiv preprint arXiv:1803.10988 (2018)
25. Ishibashi, M., Okuwa, M., Doi, S., et al.: Indices for characterizing driving style and their relevance to car following behavior. In: SICE Conference (2007)
26. Ambros, J., Turek, R., Paukrt, J.: Road safety evaluation using traffic conflicts: pilot comparison of micro-simulation and observation. In: International Conference on Traffic and Transport Engineering (2014)
27. Kitajima, S., Marumo, Y., Hiraoka, T., et al.: Comparison of evaluation features concerning estimation of driver's risk perception. Trans. Soc. Automot. Eng. Jpn. **40**(2), 191–198 (2009)
28. Zhang, T., Gao, K.: Will autonomous vehicles improve traffic efficiency and safety in urban road bottlenecks? The penetration rate matters. In: 2020 IEEE 5th International Conference on Intelligent Transportation Engineering (ICITE), pp. 366–370. IEEE (2020)
29. Yu, Y., Jiang, R., Qu, X.: A modified full velocity difference model with acceleration and deceleration confinement: calibrations, validations, and scenario analyses. IEEE Intell. Transp. Syst. Mag. (2019)

An Online Processing Method for the Cooperative Control of Connected and Automated Vehicle Platoons

Xiangyu Kong, Jiaming Wu, and Xiaobo Qu

Abstract The recent development of connected and autonomous vehicles (CAVs) makes it increasingly realistic to develop the next generation of transportation systems with the potential to improve operational performance and flexibility. The cooperative control of CAV platoons remains one of the most crucial yet challenging problems before the CAVs can be widely implemented in practice. The present study focuses on an application of CAVs at signalized intersections to realize a well-organized CAV permutation as well as improving the performance of the intersection. An online processing A* (OPA*) algorithm is developed to improve the optimality and computation performance. A comparative analysis between the proposed OPA* algorithm and an existing A* method is made. In summary, the OPA* could result in stable and scalable results which makes it possible for widely industrial usage.

Keywords Sorting · Hash table · Heuristic algorithm · Online processing A*

1 Introduction

Along with the fast development of urbanization, the transportation system also encounters unprecedented challenges, especially in cities and metropolitan areas. The traffic congestion has also been a huge problem worldwide and leads to a 272 h lose and a $2291 cost per car at most [1]. Many approaches exist for improving traffic capacity. In general, there are mainly two widely applied approaches, either extending the road network physically or introducing new technologies for a better use of traffic facilities [2–4].

The devoted efforts of improving traffic capacity in urban city mainly focus on the intersections. To enlarge the capacity of intersections, several approaches are proposed [5, 6]. Among them, Dresner and Stone found that reservation-based system or Autonomous Intersection Management (AIM) could improve the traffic performance of intersections comparing to the conventional traffic lights [7]. Matthew et al.

X. Kong · J. Wu (✉) · X. Qu
Department of Architecture and Civil Engineering, Chalmers University of Technology, 41296
Gothenburg, Sweden
e-mail: jiaming.wu@chalmers.se

© The Author(s), under exclusive license to Springer Nature Singapore Pte Ltd. 2021   133
X. Qu et al. (eds.), *Smart Transportation Systems 2021*, Smart Innovation,
Systems and Technologies 231, https://doi.org/10.1007/978-981-16-2324-0_14

illustrate the multi-intersection AIM using time-based A star which have a significant improving effect on traffic efficiency [8].

Most cases, the traffic-light-based methods are better than the reservation-based intersection methods, and the main contribution for the time reduction mentioned in the reservation method is the shorten headway, which also causes a lot of problems [9]. The theory limit of improvement for the intersection comparing to the rules nowadays can reach up to 2 times according to Sun's research [10].

Besides increasing traffic capacity, the road safety also witnesses an apparent improvement as self-driving excludes human drivers, which are one of the main contributors for road accidents [11]. In this research, each vehicle will be regarded as a tile in the grid world to find the optimal trajectories.

2 Methods

To solve the CAVs' sorting problem several approaches are investigated. Among these, the extensions of the A* algorithm show the great potential of getting solutions. Jump Point shown in the Searching (JPS) algorithm is an extended version of the A* algorithm with the help of the online pruning [12]. JPS could improve the efficiency of the A* algorithm to several orders of magnitude [13]. Besides the JPS algorithm, preprocessing algorithms are used commonly in path-finding research. One of these is called ALT (A*, Landmark, Triangle Inequality) algorithm, and it could improve the performance to more than one order of magnitude [14]. Based on the shortcuts, the contraction hierarchies (CH) algorithm is promoted [9] before the hub-based labeling (HL). HL could be six orders of magnitude faster than Dijkstra's algorithm for random (long-range) queries [15]. However, for the increasing search space, the Dijkstra algorithm is faster than the preprocessing algorithms with five orders of magnitude [16].

Moreover, the abstract, hidden, dynamic, asymmetric sorting environment has an infinite searching space. All methods above show better performance in the real world but not suitable for the infinite decision-making process. An alternative approach focuses on the model-free RL (Reinforcement Learning) method as it always shows good performance with expensive data gathering [17]. Most RL algorithms are derived from Markov Decision Process (MDP) [18]. A smart agent with the knowledge of the relationship between states and actions (value-based) for sorting will be the answer for solving the problem.

Monte Carlo Tree Search (MCTS) could also solve the N-puzzle-like problem. And the procedure between MCTS and heuristic algorithm shows a high similarity. For example, Rubik's Cube also has an extensive configuration to explore. With 44 h training, the method of the RL and MCTS could perform a 10-min median solve time comparing with the one using professional knowledge (within 1 s) [19]. By modifying the strategy of selection and backpropagation, the method Single-Player Monte-Carlo Tree Search (SP-MCTS) could solve problems similar to A* algorithm, and it could perform better when the environment does not have a specific and accurate, and appropriate evaluation function [16, 20–22]. However, few results support SP-MCTS could perform better than A* when the heuristic function is given, in this essay, the Menhaden distance or norm 1. In the Alpha Go, only the rollout itself could score 1457 with a time cost of three milliseconds. This demonstrates the traditional algorithm could improve the performance of the RL.

However, this decision-making problem is memorable which will significantly increase the difficulty of converging, not to mention getting an adorable agent. At the same time, the judge criteria in the SP-MCTS is not fixed because of the mutual transformation between good and bad actions. In this case, the modification of the A* algorithm is required for getting suitable solutions.

2.1 Improvement of A* Algorithm

The maintenance of the open list and the close list will repeatedly do operations like adding a node, selecting a node, and deleting a node. Traditional data structures like adjacency matrix or adjacency linked list sharing an average time complexity of $O(n^2)$, where n is the scale of nodes. In comparison, the binary heap could promote it to the level of $O(n \log n)$ and hash table could perform even better, with the time complexity of $O(1)$. Because of the better performance hash table shows in the domain of time consumption, in this essay, the A* algorithm will use it as the data structure for storing global information.

2.2 OPA* Algorithm

As there is no solution for a large scale sorting problem yet, finding a feasible solution has a higher priority than figuring out the optimal global solution. Based on this standard, this paper proposes the OPA* algorithm as A* algorithm could give a general solution with no restriction but not for a large number. In this case, the adding limitations should narrow the scope of the problem. To fulfill the aim mentioned before, one prerequisite is made: the maximum moving forward distance limits to 2 rows for cars on the front part. By doing this, the sub-state is limited to 3 rows for the front part, and for the last 6 cars, no matter how many rows they take, they will be sorted together to improve the global optimality.

This limitation is determined by the computing power as in the worst case, the puzzle is limiting to the sorting of six cars in six rows, and the corresponding searching space is:

$$A_{12}^6 = 13366080$$

This searching space is probably within the capacity of the A* algorithm. With the increasing computing power, moving forward for four rows could also be possible.

With the help of this prerequisite, the termination could be achieved. However, the one using OPA* needs to adjust the aim state to satisfying the assumption, which could lower down the overall optimality to a certain extent.

2.2.1 Pseudocode for the OPA* Algorithm

The start and end states are generated randomly according to the assumption for 1000 times. Then, the pseudocode for the OPA* Algorithm is shown in Table 1.

By dividing the whole platoon into several sub-platoons, the problem shrinks to the scale the A* algorithm could solve. Then, the time complexity will witness a linear increase with the increasing scale of cars. As a result, the problem of the dimensional explosion is avoided with the limitation on the dimension.

3 Results

3.1 Optimality of OPA*

In this research, the sorting steps are summarized into aggressive sorting steps. An aggressive step means all cars could move at the same time without collision. Figure 1 is performed on the 9 cars scenario with the same 100 random seeds.

From Fig. 1, the median length of decision-paths planned by the OPA* algorithm is around 8, comparing with the global optimal length given by the A* algorithm which ranging from 6 to 7. It indicates that the OPA* algorithm could be a practical and time-efficient solution for solving this kind of problem.

3.2 Generalization of OPA*

Figure 1b shows the relationship between the median value from time consumption and sorting steps against the scale of the vehicle platoon for 1000 random seeds on each scale. The color temperature represents the scale of the vehicle platoon. The larger the platoon is, the colder the color temperature will be.

A near-linear relationship is shown with some fluctuation in the time consuming, and the R square equals 0.98 with MSE (Mean Squared Error) equals to 0.0000083, as for the sorting steps, the linear relationship is even better with 1.0 (R square) and 0.19 MSE. The overall performance of the OPA* algorithm is stable, scalable, and reliable with little time consumption.

Table 1 Pseudocode for OPA* algorithm

Online processing A* algorithm	
Input: startS: The start state for the sorting problem	
endS: The aim state for the sorting problem	
sortVeh: The number of cars that have been sorted	
tailVeh: The number of cars that have not been sorted	
subStart: The start state for each sub state	
subEnd: The end state for each sub state	
carNum: The total number of vehicles	
Output:	
subClose: The optimal global results from subStart to subEnd using A*	
closeDict: A dictionary that contains global solution for solving the puzzle	
1	**def aStar**():
2	return subClose;
3	**Initialize**
4	carNum = 99; loopTime = 1;
5	subStart = startS[:9];
6	subEnd = endS[:3];
7	**while true**
8	tail = **deepcopy**(subStart);
9	**for** i **in** subEnd:
10	tail = tail.**remove**(i);
11	subClose = **aStar**(subStart, subEnd);
12	carNum - = 1;
13	**end**
14	**if** carNum == 0
15	**break;**
16	**elif** carNum == 6
17	subStart = startS[:];
18	subEnd = endS[3 * loopTime:(loopTime + 2) * 3];
19	**else**
20	subStart = startS[:9];
21	subEnd = endS[3 * loopTime:(loopTime + 1) * 3];
22	**end**

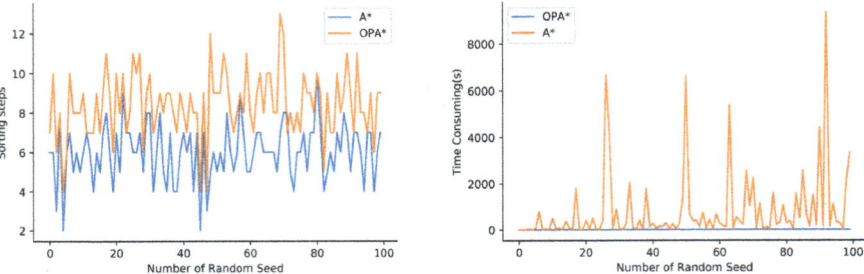

Fig. 1 Performance between A* and OPA* (a) sorting steps (b) computation time

4   Conclusion

In this article, the path planning is structured as an NP-hard path planning problem, for which A* algorithm and its extensions are proved to be useful for speeding up the process. However, A* related algorithms suffer for a nodes-unknown environment, and incomplete preprocessing decreases the performance. This essay uses hash tables for reducing the operation time complexity to around $O(1)$ if no hash crash happens but the time complexity for getting a solution still expands exponentially with the increasing dimension.

The exponential expansion of the searching space consists of both the increase of the width and the depth of the searching tree. Either reducing the width or limiting the depth or together could compensate the computing speed with the cost of the reducing optimality.

The OPA* algorithm is promoted with a hypothesis, that is, in the sorting of the subspace, a car cannot move forward or backward more than three rows. Under this assumption, the OPA* algorithm is tested with randomly generated start states and end states. In the large-scale scenario, the OPA* algorithm shows fast, scalable, and stable performance, and has the migrating ability for other similar problems for unmanned driving equipment in specific environments.

References

1. Cookson, G., Pishue, B.: Inrix Global Traffic Scorecard–Appendices. INRIX Research (2017)
2. Biswas, D., Su, H., Wang, C., Stevanovic, A., Wang, W.: An automatic traffic density estimation using single shot detection (SSD) and MobileNet-SSD. Phys. Chem. Earth Parts A/B/C **110**, 176–184 (2019)
3. Xu, Y., Zheng, Y., Yang, Y.: On the movement simulations of electric vehicles: a behavioral model-based approach. Appl. Energy **283**, 116356 (2021)
4. Gao, K., Yang, Y., Sun, L., Qu, X.: Revealing psychological inertia in mode shift behavior and its quantitative influences on commuting trips. Transp. Res. Part F Traffic Psychol. Behav. **71**, 272–287 (2020)
5. Wu, J., Liu, P., Tian, Z.Z., Xu, C.: Operational analysis of the contraflow left-turn lane design at signalized intersections in China. Transp. Res. Part C Emerg. Technol. **69**, 228–241 (2016)

6. Wu, J., Liu, P., Zhou, Y., Yu, H.: Stationary condition based performance analysis of the contraflow left-turn lane design considering the influence of the upstream intersection. Transp. Res. Part C Emerg. Technol. **122**, 102919 (2021)
7. Dresner, K., Stone, P.: Multiagent traffic management: a reservation-based intersection control mechanism. In: International Joint Conference on Autonomous Agents and Multiagent Systems, July 2004, vol. 3, pp. 530–537. IEEE Computer Society (2004)
8. Hauskknecht, M., Au, T.C., Stone, P.: Autonomous intersection management: multi-intersection optimization. In: 2011 IEEE/RSJ International Conference on Intelligent Robots and Systems, Sept 2011, pp. 4581–4586. IEEE (2011)
9. Bast, H., Delling, D., Goldberg, A., Müller-Hannemann, M., Pajor, T., Sanders, P., Wagner, D., Werneck, R.F.: Route planning in transportation networks. In: Algorithm Engineering, pp. 19–80. Springer, Cham (2016)
10. Sun, W., Zheng, J., Liu, H.X.: A capacity maximization scheme for intersection management with automated vehicles. Transp. Res. Procedia **23**, 121–136 (2017)
11. Guanetti, J., Kim, Y., Borrelli, F.: Control of connected and automated vehicles: state of the art and future challenges. Annu. Rev. Control **45**, 18–40 (2018)
12. Harabor, D., Grastien, A.: Online graph pruning for pathfinding on grid maps. In: Proceedings of the AAAI Conference on Artificial Intelligence, Aug 2011, vol. 25, no. 1
13. Harabor, D., Grastien, A.: Improving jump point search. In: Proceedings of the International Conference on Automated Planning and Scheduling, May 2014, vol. 24, no. 1
14. Goldberg, A.V., Harrelson, C.: Computing the shortest path: a search meets graph theory. In: SODA, Jan 2005, vol. 5, pp. 156–165
15. Abraham, I., Delling, D., Goldberg, A.V., Werneck, R.F.: A hub-based labeling algorithm for shortest paths in road networks. In: International Symposium on Experimental Algorithms, May 2011, pp. 230–241. Springer, Berlin, Heidelberg (2011)
16. Schadd, M.P., Winands, M.H., Van Den Herik, H.J., Chaslot, G.M.B., Uiterwijk, J.W.: Single-player Monte-Carlo tree search. In: International Conference on Computers and Games, Sept 2008, pp. 1–12. Springer, Berlin, Heidelberg (2008)
17. Clavera, I., Rothfuss, J., Schulman, J., Fujita, Y., Asfour, T., Abbeel, P.: Model-based reinforcement learning via meta-policy optimization. In: Conference on Robot Learning, Oct 2018, pp. 617–629. PMLR (2018)
18. Van Otterlo, M., Wiering, M.: Reinforcement learning and Markov decision processes. In: Reinforcement Learning, pp. 3–42. Springer, Berlin, Heidelberg (2012)
19. McAleer, S., Agostinelli, F., Shmakov, A., Baldi, P.: Solving the rubik's cube without human knowledge. arXiv preprint arXiv:1805.07470 (2018)
20. Levin, M.W., Boyles, S.D., Patel, R.: Paradoxes of reservation-based intersection controls in traffic networks. Transp. Res. Part A Policy Pract. **90**, 14–25 (2016)
21. Wu, J., Ahn, S., Zhou, Y., Liu, P., Qu, X.: The cooperative sorting strategy for connected and automated vehicle platoons. Transp. Res. Part C Emerg. Technol. **123**, 102986 (2021)
22. Wu, J., Kulcsár, B., Ahn, S., Qu, X.: Emergency vehicle lane pre-clearing: from microscopic cooperation to routing decision making. Transp. Res. Part B Methodol. **141**, 223–239 (2020)

Modeling Measurements Towards Effect of Past Behavior on Travel Behavior

Kun Gao⑩**, Tianshu Zhang, and Zhihan Li**

Abstract The inertia effect of past behavior has attracted attention in the travel behavioral literature because of its bearing on travel choice modeling. Several measurements have been proposed to model the inertia effects. However, no consensus concerning appropriate modelling methods is reached, which leads to potential biases in analysis. The study aims to conduct a comprehensive investigation of modeling measurements regarding inertia effects of past behavior from the perspectives of estimation, behavioral indications and predictions. Differing from existing literature that only focused on estimation performance, we examine the performances of different methods in predictions and behavioral interpretations. To our best knowledge, these aspects are not investigated in the literature based on empirical data. The necessary information for constructing the measurements, underlying consumption, significance in estimation, behaviorally implausible issue, performances in estimation and predictions for these measurements are all compared based on behavioral data. The results shed lights on performances and suitability of different measurements for inertia effects in terms of estimation, behavioral interpretation and prediction, which support the further investigations of past behavior on travelers' choice behavior.

Keywords Travel demand · Sustainable transportation · Sustainable behavior · Past behavior

1 Introduction

The inertia effect of past behavior has attracted attention in the travel behavioral literature because of its bearing on travel choice modeling and derivative impacts

K. Gao (✉) · T. Zhang
Architecture and Civil Engineering, Chalmers University of Technology, 412 96 Goteborg, Sweden
e-mail: gkun@chalmers.se

Z. Li
College of Arts and Sciences, University of Washington, Seattle, WA 98195, USA

X. Qu et al. (eds.), *Smart Transportation Systems 2021*, Smart Innovation, Systems and Technologies 231, https://doi.org/10.1007/978-981-16-2324-0_15

on travel demand forecasting [1–4]. In the context of transportation travel behavior research, the inertia effect of past behavior presents the impacts of past behavior on the current decision-making process and could appear as the tendency to stick with former choice or the willingness to change to a new alternative [1–3, 5–9]. Given the fact that the inertia effect is obviously related to the past experience and behavior, the modeling framework considering the inertia effects congenitally requires panel data across different periods. One of the noticeable problems in modeling the inertia effect is determining an appropriate measurement of the inertia effect that has theoretical interpretations and could be compatible with quantitative transportation models. In the context of discrete choice models, the inertia effect of past behavior is commonly represented by specific terms depending on the attributes of the alternatives in the past. Several measurements have been proposed to properly model the inertia effects in the literature [10]. The various indicators can be summarized into three main categories: the dependence dummy variable or frequency [11], measures related to the level-of-service variables of previous choice [2] and measurements depending on the differences in level-of-service variables [1, 2, 5, 12]. Different studies have used different measurements according to the available information from their data and few investigations have been conducted for the comparatively better measurement of inertia effects. These result in no consensus concerning appropriate modelling methods of inertia effect of past behavior, which further lead to potential biases in travel behavior analysis.

This study standing on the deficiency of investigating the performances of different measurements for inertia effects, comprehensively compares the performances of different measurements in terms of estimation, behavioral interpretation and prediction. Differing from existing literature that only focused on estimation performance, we comprehensively examine the performances of different methods in predictions and behavioral interpretations. To our best knowledge, these aspects are not investigated in the literature based on empirical data. The results are expected to deepen the understanding of properly modelling the inertia effect of past behavior for supporting further modeling formulation of travel choice models, which are crucial components of travel demand forecasting and transport planning.

2   Data Description

The data used in this study were reference preference (RP) and stated preference (SP) mixed datasets regarding commuting mode shift behavior collected in Shanghai, China in 2017. The respondents were asked to report their RP information about the current commuting trip, such as the most commonly used and other available transport modes, level-of-service variables (e.g. travel time, overall cost, commuting distance and comfort) and personal demographic attributes. Afterwards, based on the individual RP information, individual-specific SP scenarios were generated immediately using web-scripted code and given to the same respondent. The SP scenario was a choice between current commuting transport mode and a new hypothetical alternative. Four transport mode (private car, metro, bus and taxi) were included. The details about the survey design are available in Gao et al. [6]. Investigators were recruited to conduct face-to-face and one-to-one surveys to guarantee the validity and representativeness of data. Over 1200 respondents were initially interviewed. However, the measurements of inertia effects related to differences in level-of-service variables require full information yielding available alternatives in RP contexts. Therefore, some valid respondents have to be deleted due to lacking corresponding RP information. For instance, a car user finished the SP scenarios of car versus bus, but there were no feasible bus lines from the traveler's home to the workplace in actual RP context, which results in the impossibility to make use of it. Finally, a sample of 779 respondents (3116 efficient SP scenarios with corresponding 779 RP observations), was collected for this study. The sample percentages for private car users, metro users and bus users are 49%, 36% and 15%, respectively.

3 Model Formulation and Estimation

The model specification is built on the random utility theory depending on level-of-service (LOS) variables, inertia effects and unobservable components. The utility of alternative j perceived by individual q in SP and RP situations could be expressed by:

$$U_{jq}^{SP} = V_{jq}^{SP} + d_{jq}\phi(I_{jq}^{SP}) + \varepsilon_{jq}^{SP} \qquad \varepsilon_{jq}^{SP} \sim (0, \sigma_{\varepsilon^{SP}}^2) \tag{1}$$

$$U_{jq}^{RP} = V_{jq}^{RP} + \varepsilon_{jq}^{RP} \qquad \varepsilon_{jq}^{RP} \sim (0, \sigma_{\varepsilon^{RP}}^2) \tag{2}$$

$$\phi(I_{jq}^{SP}) = (\beta_{Ij}^{SP} + \sigma_{Ijq}^{SP}) \times I_{jq}^{SP} \tag{3}$$

$$V_{jq}^{SP} = ASC_j^{SP} + \underbrace{(\beta_{mj}^{SP} + \sigma_{mjq}^{SP}) \times LOS_{mj}^{SP}}_{DV_{jq}^{SP}} \tag{4}$$

$$V_{jq}^{RP} = ASC_j^{RP} + \underbrace{(\beta_{mj}^{RP} + \sigma_{mjq}^{RP}) \times LOS_{mj}^{RP}}_{DV_{jq}^{RP}} \tag{5}$$

where V_{jq}^{SP} and V_{jq}^{RP} are the determined segments of utility in SP and RP scenarios. DV_{jq}^{SP} and DV_{jq}^{RP} are the utility without alternative-specific constants. $\phi(I_{jq}^{SP})$ represents the inertia effect. d_{jq} is an index that equals one if $\phi(I_{jq}^{SP})$ appears in the utility function of alternative j, and zero otherwise. If the value of $\phi(I_{jq}^{SP})$ is positive, it denotes the resistance to change; if it is negative, it implies a disposition to change.

The ε_{jq}^{SP} and ε_{jq}^{RP} are the error terms with different variances (i.e., $\sigma_{\varepsilon^{SP}}^2 \neq \sigma_{\varepsilon^{RP}}^2$) in SP and RP specifications, respectively. Equation (3) assumes that the inertia effect randomly varies across individuals. β_{jq}^{SP} is the mean for inertia effect that might vary over alternative j but fixed over individuals and σ_{Ijq}^{SP} is the deviation normally distributed variances. I_{jq}^{SP} is the measurement of inertia effects. As discussed in the introduction, several measures have been proposed in the literature. In this paper, we empirically compare the most popular measurements as shown in Eq. (6) to examine their serviceability.

$$I_{jq}^{SP} = \begin{cases} Dummy\ variable \\ V_{jq}^{RP} \\ (DV_{jq}^{RP} - DV_{kq}^{RP}) & k \neq j \\ (V_{jq}^{RP} - V_{kq}^{RP}) & k \neq j \end{cases} \tag{6}$$

The first one is the state-dependence dummy variable, which is the most typical measure of inertia effect [13]. It would equal to 1 if the alternative j was chosen in the past and zero otherwise. The second one belongs to the measurements associated with level-of-service variables and counted by the utility of alternative j in the RP situation [2]. Herein, we do not test the measurements using only one attribute of past choice (e.g. RP travel time). The utility of previously used option is determined by all level-of-service variables rather than one attribute. It is not logical to ignore other influencing factors according to the basic thinking of the measurements. Using the RP travel time for the measurement of inertia effect is a special case of V_{jq}^{RP} where the utility only depends on travel time. The third measurement in Eq. (6) was proposed by Cantillo et al. [1] and calculated by the difference in the level-of-service variables of alternative j and other alternative k in the RP situation. The fourth measurement is similar to the third measurement, but includes the alternative-specific constants to consider unobserved preferences for different modes. It should be noted that specific inertia effect terms for different modes are used in this study.

We apply joint RP/SP estimation to make full use of behavioral information collected from the data. The joint estimation of RP and SP datasets needs scaling the utility in the SP situations to make the variances in both datasets equivalent [14]. Therefore, the final utility functions can be expressed as:

$$U_{jq}^{SP} = \lambda_{SP}\left(V_{jq}^{SP} + d_{jq}\phi(I_{jq}^{SP}) + \varepsilon_{jq}^{SP}\right) \tag{7}$$

$$U_{jq}^{RP} = \lambda_{RP}\left(V_{jq}^{RP} + \varepsilon_{jq}^{RP}\right) \tag{8}$$

where $\lambda_{SP}\varepsilon_{jq}^{SP} = \lambda_{RP}\varepsilon_{jq}^{RP}$. The scale parameter is the only control parameter for obtaining the same variance in the RP and SP datasets [2, 13, 14]. We normalize the model based on the RP data (i.e., $\lambda_{RP} = 1$). The joint estimation of the RP and SP datasets is conducted by simulated maximum likelihood estimation methods

[15]. To examine the performances of the selected measurements of inertia effect in both estimation and prediction, we divide the overall data into calibration part and validation part [9, 16, 17]. 85% of all respondents were randomly extracted to be estimation dataset and the remaining 15% respondents were prediction dataset. The above models were calibrated based on the estimation dataset for testing estimation performances and then applied into the prediction datasets for testing predictive performance. To diminish the possible random effects of selecting estimation and prediction datasets, the above process was repeated five times, referring to the cross-validation methods in machine learning. Therefore, the performances in estimation and prediction were measured by the overall likelihood of the sample. To avoid the unexpected signs of parameters in the estimation, one-sided constrained triangular distribution were used to constrain the signs of estimated parameters for cost, travel time and crowding levels [18]. Pythonbiogeme was applied to estimate the models with 1000 random draws. The panel effect was considered and the error component models were performed firstly to examine several possible nested structures among modes and serial correlations among RP and SP situations.

4 Results and Analysis

4.1 Performances in Estimation and Behavioral Interpretation

The estimated results of estimation datasets are demonstrated in Table 1. We actually generated five sets of data for estimation. For the sake of the text limit, we merely show one sample since the other four estimation results reveal similar principles. On overall goodness of fit in terms of the log-likelihood, Akaike Information Criterion and Bayesian Information Criterion, the Model 1 and Model 4 are comparatively preferred followed by Model 3 and Model 2 consecutively. The differences in model fitness are significant on the standard likelihood ratio test. However, the differences in terms of model fitness are not very substantial. Besides the estimation fitness, it is more crucial to analyze the behavioral hints from the results to examine if they are in accordance with their behavioral assumption and intuition.

For Model 1 using dependence dummy variables, most of the estimated coefficients are significant at 95% confidence level. The estimated inertia effect of car is up to 10.6 with a standard deviation (SD) of 0.939. The mean inertia effect of the car users is equivalent to the incremental utility of 30 CNY reduction (1 CNY = 0.15$) in cost. It indicates that most car users show strong resistance to change in mode shift and are unwilling to switch from private cars to others. The inertia term for metro is 0.241, but cannot be rejected to be different from 0. The SD of the inertia term for metro is 2.93 and significant (t-value: 2.89) indicating large variances exist among metro users' inertia effect due to past behavior. It can be deduced from the results some metro users have positive inertia effect (i.e. resistance to change) while others

Table 1 Estimated results

Parameters	Model 1 Dependence dummy variable		Model 2 RP utility		Model 3 RP utility difference without ASC		Model 4 RP utility difference with ASC	
	Value	t-stat.	Value	t-stat.	Value	t-stat.	Value	t-stat.
Cost (mean)	−0.353	−4.68	−0.357	−3.93	−0.317	−3.41	−0.363	−4.78
Cost (st. dev.)	0.353	4.68	0.357	3.93	0.317	3.41	0.363	4.78
Travel time_car_SP (mean)	−0.143	−3.13	−0.276	−3.17	−0.174	−2.63	−0.147	−3.52
Travel time_car_SP (st. dev.)	0.143	3.13	0.276	3.17	0.174	2.63	0.147	−3.52
Travel time_PT_SP (mean)	−0.159	−4.09	−0.133	−3.35	−0.152	−3.16	−0.120	−3.97
Travel time_PT_SP (st. dev.)	0.159	4.09	0.133	3.35	0.152	3.16	0.120	3.97
Crowd2_metro_SP (mean)	−3.91	−3.99	−3.49	−3.45	−3.21	−2.91	−2.52	−3.40
Crowd2_metro_SP (st. dev.)	3.91	3.99	3.49	3.45	3.21	2.91	2.52	3.40
Crowd3_metro_SP (mean)	−6.76	−4.12	−5.80	−3.66	−5.66	−3.35	−4.89	−4.13
Crowd3_metro_SP (st. dev.)	6.76	4.12	5.80	3.66	5.66	3.35	4.89	4.13
Crowd2_bus_SP (mean)	−3.18	−3.62	−3.30	−3.50	−2.22	−2.35	−2.11	−3.28
Crowd2_bus_SP (st. dev.)	3.18	3.62	3.30	3.50	2.22	2.35	2.11	3.28

(continued)

Table 1 (continued)

Parameters	Model 1 Dependence dummy variable		Model 2 RP utility		Model 3 RP utility difference without ASC		Model 4 RP utility difference with ASC	
	Value	t-stat.	Value	t-stat.	Value	t-stat.	Value	t-stat.
Crowd3_bus_SP (mean)	−3.94	−3.35	−3.67	−3.03	−3.58	−2.85	−3.15	−3.33
Crowd3_bus_SP (st. dev.)	3.94	3.35	3.67	3.03	3.58	2.85	3.15	3.33
Error component metro and bus	6.99	4.49	6.05	3.28	4.69	2.57	4.15	3.48
ASC_car_SP	−1.63	−1.05	6.52	2.96	3.67	2.37	0.0701	0.06
ASC_metro_SP	5.12	3.71	4.91	3.36	4.61	3.55	2.71	3.02
Travel time_car_RP (mean)	−0.036	−2.28	−0.0239	−1.71	−0.0468	−3.07	−0.0596	−3.77
Travel time_car_RP (st. dev.)	0.036	2.28	0.0239	1.71	0.0468	3.07	0.0596	3.77
Travel time_PT_RP (mean)	−0.129	−3.95	−0.188	−3.49	−0.147	−2.79	−0.148	−3.22
Travel time_PT_RP (st. dev.)	0.129	3.95	0.188	3.49	0.147	2.79	0.148	3.22
Crowd2_metro_RP (mean)	−0.0917	−0.12	−0.423	−0.41	−2.01	−2.34	−0.106	−0.106
Crowd2_metro_RP (st. dev.)	0.0917	0.12	0.423	0.41	2.01	2.34	0.106	0.06

(continued)

Table 1 (continued)

Parameters	Model 1 Dependence dummy variable		Model 2 RP utility		Model 3 RP utility difference without ASC		Model 4 RP utility difference with ASC	
	Value	t-stat.	Value	t-stat.	Value	t-stat.	Value	t-stat.
Crowd3_metro_RP (mean)	−1.72	−1.75	−1.80	−1.53	−3.51	−3.31	−0.979	−0.97
Crowd3_metro_RP (st. dev.)	1.72	1.75	1.80	1.53	3.51	3.31	0.979	0.97
Crowd2_bus_RP (mean)	−1.50	−3.35	−3.20	−3.33	−2.74	−1.76	−1.51	−2.50
Crowd2_bus_RP (st. dev.)	1.50	3.35	3.20	3.33	2.74	1.76	1.51	2.50
Crowd3_bus_RP (mean)	−2.32	−3.62	−4.28	−3.70	−3.32	−2.53	−1.70	−2.65
Crowd3_bus_RP (st. dev.)	2.32	3.62	4.28	3.70	3.32	2.53	1.70	2.65
ASC_car_RP	3.20	2.78	−0.0505	−0.06	4.96	2.49	3.08	3.83
ASC_metro_RP	1.07	3.71	−0.707	−0.73	1.10	3.17	0.474	0.28
Inertia_car (mean)	10.6	3.91	−1.41	−7.63	0.848	3.16	1.38	6.40
Inertia_car (st. dev.)	0.939	0.90	−0.368	−1.93	1.12	3.08	0.604	6.45
Inertia_metro (mean)	0.241	0.24	−0.0251	−0.25	0.302	1.91	0.664	3.91
Inertia_metro (st. dev.)	2.93	2.89	0.212	1.69	0.0628	0.13	0.217	1.35
Inertia_bus (mean)	−0.309	−0.23	0.0126	0.1	0.146	0.58	0.114	0.73
Inertia_bus (st. dev.)	1.71	0.44	0.0149	0.2	0.413	1.23	0.531	2.27

(continued)

Table 1 (continued)

Parameters	Model 1 Dependence dummy variable		Model 2 RP utility		Model 3 RP utility difference without ASC		Model 4 RP utility difference with ASC	
	Value	t-stat.	Value	t-stat.	Value	t-stat.	Value	t-stat.
Scale parameter (SP to RP)	0.362	−8.20	0.337	5.63	0.409	4.79	0.333	4.33
Model fit								
Final log-likelihood	−1528.312		−1535.775		−1533.399		−1531.908	
No. of parameters	25		25		25		25	
AIC	3106.625		3121.551		3116.797		3113.815	
BIC	3258.343		3273.269		3267.51		3273.533	

Note AIC and BIC are the Akaike Information Criterion and Bayesian Information Criterion, respectively. The smaller AIC and BIC indicates a better model fit. st. dev. means standard deviation. The t-stat. in the table is robust t-value. The t-statistic corresponding to the scale parameter is computed with respect to a value of 1

show a disposition to change. For the bus users, a mean inertia term of −0.309 with an SD of 1.71. The mean value and SD for bus are not significant.

For Model 2 using the RP utility to measure the inertia effect, the estimated inertia effect for car is −1.41 on average and has an SD of 0.368. Given that the RP utility in Model 2 is always negative, the result virtually denotes car users' resistance to change in mode shift. Nevertheless, the negative inertia term means that the smaller is the RP utility is (i.e., the worse is the used option), the more stickiness the traveler shows to using private car. The result is not really in line with behavioral intuition and basic thinking of the measurement. It is expected that the better is the RP utility is, the more unwilling is the traveler to change. The mean values of inertia effects for metro and bus are both small, positive and non-significant. Substantial variances are observed in the inertia effects of bus and metro as in Model 1.

For the Model 3 and 4 using the differences in utilities as the measurement of inertia effects, the estimated results from the two models are similar, while the Model 4 performs better in terms of model fitness and significances of parameters. We can find highly significant positive inertia effect for car users with a mean of 1.38 and an SD of 0.604 from Model 4. The result means that the larger is the utility of car than other alternatives in the RP context, the more resistance to change the car users show in SP choices. It is logical and in line with basic assumption of the measurement. The mean inertia effects for metro is 0.664 and significant, indicating that the metro users show resistance to change on average. The mean value of the inertia effect for bus is very small (0.114) and not significant. However, we can find that the standard deviation of inertia term of bus is significant at a 95% confidence level and much larger than the mean value, which demonstrates significant heterogeneity among bus users.

4.2   Performances in Prediction

Using sample enumeration, we can calculate the prediction accuracy towards travelers' mode shift behavior of Model 1–4 for each individual in the validation sample. We use the log-likelihood and corresponding probability of correctly predicting the selected mode in SP scenarios as the measurement of prediction performances [19]. The larger are the log-likelihood and probability of correct prediction, the better the model performs in prediction. Paired comparisons are employed to identify the differences in prediction results from different models. The comparison results among Model 1–4 are demonstrated in Table 2. The Model 2 using RP utility as the measurement of inertia effect has larger log-likelihood and probability of correct prediction than other measurements according to results, indicating that the measurement of RP utility is comparatively superior to other measurements in prediction. The advantages of Model 2 in prediction over other models are significant at a confidence level of 95% (except the Pair 1 in the probability of correct prediction). The Model 1 using dependence dummy variable performs better in prediction compared to Model 3 and 4 who employ the differences in RP utility as the measurement of

Table 2 The results of prediction overall validation sample

		Mean	Std. deviation	Std. error mean	95% confidence interval of the difference		t	Sig. (2-tailed)
					Lower	Upper		
		Paired differences (individual predicted log-likelihood)						
Pair 1	M1 − M2	−0.0473	0.4065	0.0203	−0.0873	−0.0073	−2.327	0.020
Pair 2	M1 − M3	0.0181	0.2449	0.0122	−0.0060	0.0422	1.480	0.140
Pair 3	M1 − M4	0.0352	0.2851	0.0143	0.0072	0.0632	2.469	0.014
Pair 4	M2 − M3	0.0654	0.4132	0.0207	0.0248	0.1060	3.167	0.002
Pair 5	M2 − M4	0.0825	0.5337	0.0267	0.0300	0.1350	3.092	0.002
Pair 6	M3 − M4	0.0171	0.2163	0.0108	−0.0042	0.0383	1.579	0.115
		Paired differences (individual probability of correct prediction)						
Pair 1	M1 − M2	−0.0123	0.1736	0.0087	−0.0293	0.0048	−1.415	0.158

(continued)

Table 2 (continued)

		Mean	Std. deviation	Std. error mean	95% confidence interval of the difference		t	Sig. (2-tailed)
					Lower	Upper		
Pair 2	M1 – M3	0.0140	0.1070	0.0054	0.0034	0.0245	2.608	0.009
Pair 3	M1 – M4	0.0114	0.1152	0.0058	0.0000	0.0227	1.973	0.049
Pair 4	M2 – M3	0.0262	0.1824	0.0091	0.0083	0.0442	2.876	0.004
Pair 5	M2 – M4	0.0236	0.2106	0.0105	0.0029	0.0443	2.246	0.025
Pair 6	M3 – M4	−0.0026	0.0623	0.0031	−0.0087	0.0035	−0.830	0.407

Note M1–4 denote the Model 1–4

inertia effects. The discrepancy is statistically significant in terms of log-likelihood and correctly predicting probability. The predicting log-likelihood by Model 3 is comparatively larger than but not significantly different from Model 4. There are no obvious differences in the correct prediction probabilities between Model 3 and 4. We can conclude that the Model 3 and 4 perform similarly in predicting travelers' mode shift behavior.

According to the estimated results in Table 1, we can find that the inertia effects for metro and bus users are not significant and have noticeable variances in some models (e.g. Model 1 and 3). However, the inertia effect for car are highly significant and show resistance to change in all models. To eliminate the possible impacts of non-significant parameters, we further test the prediction power of different models based on a reduced validation sample. The reduced validation sample only contains car users. The comparison outcomes are presented in Table 3. In terms of the individual predicted log-likelihood, the Model 2 using RP utility shows significantly (at a confidence level of 95%) better predicting power in contrast to others. The probabilities of correct prediction by Model 2 are also 1.6–2.2% larger than other models. However, the differences between Model 2 with Model 1 and Model 4 in the probabilities of correct prediction are only statistically significant at the confidence level of 90%, rather than 95%. It can be summarized that Model 2 performs better but not noticeably better in the predictions for the reduced sample compared to other models. The predicted log-likelihood of Model 1 using dependence dummy variable is significantly larger than Model 4 and has no obvious distinction compared to the Model 3. However, we cannot find substantial and significant differences at a confidence level of 95% among Model 1, 3 and 4 in terms of correctly predicting probability. Model 3 has larger log-likelihood value compared to Model 4. Nevertheless, there is no substantial and significant distinction between the correct prediction probability of Model 3 and 4. We can summarize that there are no marked differences in the predicting power among Model 1, 3 and 4 according to the above results.

5 Summary

The study is aimed to conduct a comprehensive comparison of measurements regarding inertia effects of past behavior from the perspectives of estimation, behavioral indications and predictions. The overall comparisons could be summarized in Table 4. The necessary information for constructing the measurements, underlying consumption, significance in estimation, behaviorally implausible issue, performances in estimation and predictions for these measurements are all illustrated. The results shed lights on performances and suitability of different measurements for inertia effects in terms of estimation, behavioral interpretation and prediction. The results provide many implications for formulating appropriate measurement towards inertia effect of past behavior in discrete choice models and thus support the further investigations of past behavior on travelers' choice behavior.

Table 3 The results of prediction over reduced validation sample (car users)

		Mean	Std. deviation	Std. error mean	95% confidence interval of the difference		t	Sig. (2-tailed)
					Lower	Upper		
Paired differences (individual predicted log-likelihood)								
Pair 1	M1 – M2	−0.0586	0.4293	0.0248	−0.1074	−0.0098	−2.364	0.019
Pair 2	M1 – M3	0.0091	0.1208	0.0070	−0.0047	0.0228	1.3	0.195
Pair 3	M1 – M4	0.0453	0.2444	0.0141	0.0176	0.0731	3.213	0.001
Pair 4	M2 – M3	0.0677	0.4253	0.0246	0.0193	0.1160	2.756	0.006
Pair 5	M2 – M4	0.1039	0.5814	0.0336	0.0379	0.1700	3.096	0.002
Pair 6	M3 – M4	0.0363	0.2452	0.0142	0.0084	0.0641	2.561	0.011
Paired differences (individual probability of correct prediction)								
Pair 1	M1 – M2	−0.0169	0.1838	0.0106	−0.0377	0.0040	−1.588	0.113
Pair 2	M1 – M3	0.0052	0.0515	0.0030	−0.0006	0.0111	1.763	0.079
Pair 3	M1 – M4	0.0057	0.0775	0.0045	−0.0031	0.0145	1.278	0.202

(continued)

Table 3 (continued)

		Mean	Std. deviation	Std. error mean	95% confidence interval of the difference		t	Sig. (2-tailed)
					Lower	Upper		
Pair 4	M2 – M3	0.0221	0.1852	0.0107	0.0011	0.0431	2.066	0.040
Pair 5	M2 – M4	0.0226	0.2223	0.0128	–0.0027	0.0478	1.759	0.080
Pair 6	M3 – M4	0.0005	0.0609	0.0035	–0.0064	0.0074	0.138	0.891

Table 4 Summary for the comparison about different measurements of inertia effects

	Model 1 Dependence dummy variable	Model 2 RP utility		Model 3 RP utility difference without ASC	Model 4 RP utility difference with ASC
Data requirement	The previously used option	The detailed information of the previously used option		The detailed information of the previously used option and other alternatives	The detailed information of the previously used option and other alternatives
Underlying assumption	None	Inertia effect is related to the performance of the used option		Inertia effect is related to the differences in the performances of used option and other alternatives	Inertia effect is related to the differences in the performances of used option and other alternatives
Significance of estimated inertia effects in this study (95% confidence level)		Car: sig Metro: not sig Bus: not sig	Car: sig Metro: not sig Bus: not sig	Car: sig Metro: sig Bus: not sig	Car: sig Metro: sig Bus: sig
Behavioral implausible issue		*No*	*Yes*	*Partly*	*Partly*
Performance in estimation	AIC	3106.625	3121.551	3116.797	3113.815
	BIC	3258.343	3273.269	3267.515	3273.533
	Rank	1	4	3	2
Performance in prediction (full validation sample)	Average log-likelihood	−0.65326	−0.60595	−0.67138	−0.68846
	Average probability of correct prediction	0.571535	0.583812	0.557583	0.560167
	Rank	2	1	4	3
Performance in estimation (reduced validation sample)	Average log-likelihood	−0.65302	−0.59442	−0.66209	−0.69835
	Average probability of correct prediction	0.566261	0.583113	0.561024	0.560541
	Rank	2	1	3	4

References

1. Cantillo, V., Ortúzar, J.d.D., Williams, H.C.: Modeling discrete choices in the presence of inertia and serial correlation. Transp. Sci. **41**(2), 195–205 (2007)
2. Cherchi, E., Manca, F.: Accounting for inertia in modal choices: some new evidence using a RP/SP dataset. Transportation **38**(4), 679–695 (2011)
3. Cherchi, E., Cirillo, C.: Understanding variability, habit and the effect of long period activity plan in modal choices: a day to day, week to week analysis on panel data. Transportation **41**(6), 1245–1262 (2014)
4. Xu, Z., et al.: Modeling relationship between truck fuel consumption and driving behavior using data from internet of vehicles. Comput.-Aided Civ. Infrastruct. Eng. **33**(3), 209–219 (2018)
5. González, R.M., Marrero, Á.S., Cherchi, E.: Testing for inertia effect when a new tram is implemented. Transp. Res. Part A Policy Pract. 150–159 (2017)
6. Gao, K., et al.: Inertia effects of past behavior in modal shift behavior: interactions, variations and implications for demand estimation (2020)
7. Gao, K., et al.: Quantifying economic benefits from free-floating bike-sharing systems: a trip-level inference approach and city-scale analysis. Transp. Res. Part A Policy Pract. **144**, 89–103 (2021)
8. Kuang, Y., Qu, X., Wang, S.: A tree-structured crash surrogate measure for freeways. Accid. Anal. Prev. **77**, 137–148 (2015)
9. Zhao, X., et al.: Field experiments on longitudinal characteristics of human driver behavior following an autonomous vehicle. Transp. Res. Part C Emerg. Technol. **114**, 205–224 (2020)
10. Gao, K., et al.: Revealing psychological inertia in mode shift behavior and its quantitative influences on commuting trips. Transp. Res. Part F Traffic Psychol. Behav. **71**, 272–287 (2020)
11. Srinivasan, K.K., Bhargavi, P.: Longer-term changes in mode choice decisions in Chennai: a comparison between cross-sectional and dynamic models. Transportation **34**(3), 355–374 (2007)
12. Xu, Y., Zheng, Y., Yang, Y.: On the movement simulations of electric vehicles: a behavioral model-based approach. Appl. Energy 116356 (2020)
13. Adamowicz, W.L.: Habit formation and variety seeking in a discrete choice model of recreation demand. J. Agric. Resour. Econ. **19**(1), 19–31 (1994)
14. Bhat, C.R., Castelar, S.: A unified mixed logit framework for modeling revealed and stated preferences: formulation and application to congestion pricing analysis in the San Francisco Bay area. Transp. Res. Part B Methodol. **36**(7), 593–616 (2002)
15. Train, K.E.: Discrete Choice Methods with Simulation. Cambridge University Press (2009)
16. Qu, X., et al.: Jointly dampening traffic oscillations and improving energy consumption with electric, connected and automated vehicles: a reinforcement learning based approach. Appl. Energy **257**, 114030 (2020)
17. Zhou, M., Qu, X., Li, X.: A recurrent neural network based microscopic car following model to predict traffic oscillation. Transp. Res. Part C Emerg. Technol. **84**, 245–264 (2017)
18. Durán-Hormazábal, E., Tirachini, A.: Estimation of travel time variability for cars, buses, metro and door-to-door public transport trips in Santiago, Chile. Res. Transp. Econ. **59**, 26–39 (2016)
19. Vij, A., Walker, J.L.: How, when and why integrated choice and latent variable models are latently useful. Transp. Res. Part B Methodol. **90**, 192–217 (2016)

Measuring Transportation Accessibility Based on Different Data Sources: A State-of-the-Art Review

Ke Ren, Can Cui, and Yadan Yan

Abstract Accessibility plays an important role in the field of transportation. In previous studies, due to the limitation of data sources, the research on dynamic accessibility is limited. In recent years, the emergence of new data sources provides possibilities for the dynamic accessibility. This paper summarizes four classic accessibility evaluation models, including the space separation measure, cumulative opportunities measure, potential accessibility measure, and space–time prism. Moreover, this paper introduces the limitations of traditional data sources and analyzes the characteristics of new data sources such as floating car data, smart cards data, mobile phone recording data, and navigation map (API) data. A comprehensive overview of the application of different data sources in transportation accessibility is also developed. Finally, this review study shows opportunities and challenges for transportation accessibility studies.

Keywords Transportation accessibility · Big data · Review

1 Introduction

Accessibility analysis has played an important role in transportation planning. Although the concept of accessibility has been discussed in literature for more than sixty years, it is still hard to define and measure. Hansen [1] defined accessibility as the potential of opportunities for inter-action. Geurs and Jan [2] described it as the degree of easiness to which a hypothetical land use and a given transportation system allow persons or goods to reach activities or destinations by combining means of transport. Accessibility was commonly calculated on the basis of a person's ease of carrying out desirable activities, using desirable means of transport and at the desired time [3]. Bertolini et al. [4] introduced accessibility as the number and diversity of places that can be reached within a given travel time or cost. There are different measuring methods of accessibility from different perspectives, e.g.,

K. Ren · C. Cui · Y. Yan (✉)
School of Civil Engineering, Zhengzhou University, Zhengzhou 450001, P. R. China
e-mail: yanyadan@zzu.edu.cn

© The Author(s), under exclusive license to Springer Nature Singapore Pte Ltd. 2021 159
X. Qu et al. (eds.), *Smart Transportation Systems 2021*, Smart Innovation,
Systems and Technologies 231, https://doi.org/10.1007/978-981-16-2324-0_16

(1) travel impedance, considering factors of travel time, travel distance, travel cost, the degree of congestion, comfort, and safety; (2) the attractiveness of destinations, including population density, infrastructure system, population size, regional area, number of jobs, number of points of interest, and travel demand, etc. Li et al. [5] aimed to maximize the total weighted benefits between users and multiple services by studying location design. Meanwhile, with the development of science and technology, different data sources continuously emerge. How to use multi-source big data for accessibility research has attracted a lot of attention in recent years.

Many different studies on transportation accessibility have been conducted during the last decade. However, none of the previous studies present a comprehensive review of literature based on the data sources. Neilson et al. presented an extensive review of published research that measures active accessibility [6]. Some reviews were conducted from the perspective of transit accessibility models [7] and from the perspective of temporal dimension of accessibility [8]. Some experts and scholars have also reviewed and analyzed the application of big data in the transportation field [9–11]. But the research scope is wide and the application and challenges of big data in transportation accessibility research have not been mentioned. Therefore, this study focuses on the review of multi-source big data in transportation accessibility research, with the purpose of suggesting implications for future accessibility research.

2 Accessibility Measure

Four perspectives on measuring accessibility are identified in existing studies, and they are infrastructure-based measures, location-based measures, person-based measures, and utility-based measures, respectively [12]. The four classic models for transportation accessibility measurement are summarized as follows.

2.1 Space Separation Measure

The accessibility model based on space separation defines accessibility as the difficulty of overcoming the space separation between two points. The index of space separation can be expressed by distance, time and cost etc., shown as follows:

$$A_i = \frac{1}{n} \sum_{j=1, j \neq i}^{n} d_{ij} \tag{1}$$

where A_i represents the accessibility of zone i; n is the number of zones; d_{ij} is the impedance for traveling between zone i and j, generally expressed by travel distance.

The model reflects the spatial position relationship between nodes, which is simple in form and easy to calculate. But it focuses on the transportation network itself without considering many factors such as land use and travel demand, and lacks the embodiment of the true connotation of accessibility.

2.2 Cumulative Opportunities Measure

The cumulative opportunities measure is also called the isochrones measure. It calculates the number of service opportunities accessible to a certain location under a specified travel cost (e.g., distance, time, and expense). The more opportunities, the higher the level of accessibility. The basic formula [12] is as follows:

$$A_i = \int_0^T f(t)dt \tag{2}$$

where T represents the chosen travel cost; $f(t)$ is the opportunity distribution function that varies with travel costs.

Since the cumulative opportunity method does not make assumptions about passenger travel experience, and land use, this method requires less data to calculate accessibility. It is easy to explain the physical meaning of accessibility, but it fails to consider travel distance. The impact of attenuation is that all opportunities within the limit of travel costs can be obtained by waiting for opportunities.

2.3 Potential Accessibility Measure

Potential accessibility measure believes that accessibility refers to the degree of difficulty for a place to obtain services. This degree of difficulty is inversely proportional to the spatial distance between the demand point and the service point, and is directly proportional to the service capacity or scale of the service point.

Hansen [1] proposed a potential model to measure accessibility in 1959. Later, some scholars improved it and proposed a gravity model, combining the ideas of space obstruction and opportunity accumulation and drawing on the law of gravity in physics.

$$A_i = \sum_{j=1}^n \frac{M_j}{f(d_{ij})} \tag{3}$$

where M_j indicates the service capability of the service point j; $f(d_{ij})$ is the impedance function; d_{ij} represents the traffic impedance between the demand point i and the service point j, usually expressed by spatial distance.

2.4 Space–Time Prism

The space–time prism model is also called the individual characteristic model, which is used to calculate the accessibility of an individual for a specific opportunity within a given period of the day. From the perspective of space–time geography, accessibility refers to the use of space–time prisms to describe the time–space range that an individual can reach and the corresponding activity opportunities from the perspective of individual travelers under certain time–space constraints to evaluate characteristics of individual travel activities. The model can intuitively display the time and space constraints of travel individuals and the corresponding possible activity spaces. Based on different constraints and activity spaces, it reflects the diversity of individual travel behaviors, but the model needs a large amount of data to support. The basic formula is as follows:

$$A = \sum_i W_i I(i) \tag{4}$$

$$I(i) = \begin{cases} 1, & \text{if } i \in FOS \\ 0, & \text{otherwise} \end{cases} \tag{5}$$

where W_i represents the number of opportunities in zone i; FOS is the feasible opportunity set.

3 Application of Big Data in Measuring Transportation Accessibility

3.1 Data Description

Traditional data sources mainly include field survey data and static sensor data. Static sensors are expensive, and are often concentrated on highways or high-grade roads. Field surveys are time-consuming and laborious, and unable to obtain accurate and comprehensive travel information.

New multi-source big data provides new possibilities for dynamic accessibility research. As a typical representative of big data, floating car data (FCD) is generated by taxis or buses equipped with GPS. The taxi's on-board GPS system collects the position coordinates, driving speed, and passenger status of the taxi at regular intervals, and then stores the data in the database through wireless transmission. These data are easily obtained and have the characteristics of all-weather, wide coverage, strong real-time and high information accuracy. Moreover, with the development and popularization of intelligent transportation systems, transportation smart cards are another major source of transportation data. The smart card data contain information about the user's travel mode, origin, destination, start time, and end time. It is also worth noting that the availability of mobile phone recording data opens up new opportunities for dynamic accessibility analysis. Phone data includes anonymous user identification, call time, and user location at the base station level.

Internet open data, especially map-based open platforms, provide more accurate data support for accessibility research. The advantages of Internet open data are wide coverage, fast update, and authentic data. For instance, TomTom's Speed Profiles data can be used to obtain relevant speed data. Open-access General Transit Feed Specification (GTFS) data can be used to retrieve up-to-date routes and schedules for public transport. OpenStreetMap is helpful for obtaining road network data. Navigation map APIs can be used to obtain information such as travel time, distance, cost, the number of transfers, and walking time between OD pairs.

3.2 Application of Multi-source Big Data

Combining new data with classic accessibility evaluation models has great research prospects and is expected to improve the quality, quantity and frequency of available information, thereby improving the efficiency and effectiveness of transportation and urban planning policies. Studies of the application of multi-source data in transportation accessibility research are shown in Table 1.

Wang et al. [13] constructed an inter-regional accessibility evaluation model based on taxi GPS data, and extract complete trips from taxi GPS data to calculate the actual trip time to improve the average accuracy of travel time. Jiang et al. [14] used large-scale taxi GPS data to compute real-time taxi accessibility in Beijing, China, while Cui et al. used taxi GPS data to measure network accessibility [15]. Xu et al. [16] developed two models that have stronger performed in predicting fuel consumption in new routes based on Internet of Vehicles data. Based on multi-source data such as bus smart card data and departure time interval data, Yu et al. [17] proposed a method for calculating station waiting time considering routes and traffic, and established a two-stage opportunity mode. Wu et al. [18] comprehensively used Nanjing rail transit AFC data, shared bicycle operation data, and online map POIs data to explore the connection characteristics of shared bicycles and rail transit, and established a model to calculate transit station accessibility. Chen et al. [19] provided an acquisition method for more accurate travel time data in the multimodal public transportation network on the basis of an Internet mapping service, which simplified road network modeling efforts. Guan et al. [20] proposed a new index called dynamic modal accessibility gap (DMAG) and used taxi GPS and metro smart card data as well as POIs data to generate the DMAG index. Gao et al. [21] used data collected in Shanghai of China in Autumn to examine the effects of irrational psychological inertia in mode shift behavior. Pedro et al. [22] used mobile phone records and online route planners to conduct a dynamic analysis of urban accessibility considering its two main components: travel times and the attractiveness of destinations.

It can be found that the accessibility measures are relatively mature. For a long time, transportation planners, engineers, and decision makers have been committed

Table 1 Application of multi-source big data in transportation accessibility research

Measure	Data types	Data sources	References
Cumulative opportunities measure	Travel time	Taxi GPS data	[15]
Cumulative opportunities measure	Travel time; travel distance; OD-travel-matrix	2011 census data; employed labor force commuting trips data; GTFS data	[23]
Potential accessibility measure	Travel time; opportunities	TomTom's Speed Profiles data; 2011 census data	[24]
Space separation measure	Travel time	Baidu map LBS open platform	[25]
Potential accessibility measure	The number of potential-opportunities; travel time	GTFS data	[26]
Space separation measure	Routes and schedules for PT; travel time; the distribution of population	GTFS data; OpenStreetMap road network data; population register data	[27]
Space–time prisms	Transit networks and schedules data; transit and pedestrian network data; healthcare services data	GTFS data; OpenTripPlanner tools; Infogroup business data	[28]
Cumulative opportunities measure	Travel time; OD travel matrices	Google Maps API; mobile phone records	[22]
Space separation measure	Road network data; travel time; spatial locations of O and D points	OpenStreetMap database; Baidu Maps API	[29]
Cumulative opportunities measure	Travel time; distance; the number of POIs	Taxi GPS data; metro smart card data; Baidu Maps API	[19]
Potential accessibility measure	Travel time; travel distance; the number of traffic lights	Baidu Maps path data	[30]

to study accessibility through models based on traditional static surveys and historical data. New data brought about by emerging technologies can help increase opportunities for accessibility research. Furthermore, new sources of data show good application in time-sensitive accessibility (i.e., dynamic accessibility) studies because they produce more accurate and realistic information than static or partially dynamic analysis.

4 Conclusion

This study reviews the accessibility research from the perspective of common types of new multi-sources data combined with classic accessibility measures. In terms of accessibility measures, data types, and data sources, etc., the past five-years accessibility studies are sorted out in chronological order. New data sources are suitable for various time-sensitive types of dynamic accessibility modeling, and the emergence of new data sources alleviates this limitation.

Though increasing availability of multi-sources big data provide numerous opportunities for accessibility research, these data also pose many challenges, including access permission to use such data for research, data governance, ethics, and privacy. These challenges can be summarized as data collection, quality, storage, and security issues [10]. For example, the possibility that the data from GPS-enabled smartphones or vehicle-embedded sensors share data with unconsented third parties, such as Google Maps, will raise privacy concerns. This raises issues of personal safety accruing to commercialization of personal data and possibility of data leakage. To some extent, the quality of data affects data processing and accurate interpretation.

References

1. Hansen, W.: How accessibility shapes land uses. J. Am. Inst. Plann. **10**(2), 73–76 (1959)
2. Geurs, K., Jan, R.: Accessibility Measures: Review and Applications. Rivm Report National Institute of Public Healyh & Environment, Bilthoven (2001)
3. Bhat, C., Handy, S., Kockelman, K., Mahmassani, H., Chen, Q., Weston, L.: Development of an Urban Accessibility Index: Literature Review. Center for Transportation Research, University of Texas at Austin (2000)
4. Bertolini, L., Clercq, F., Kapoen, L.: Sustainable accessibility: a conceptual framework to integrate transport and land use plan-making. Two test-applications in the Netherlands and a reflection on the way forward. Transp. Policy **12**(3), 207–220 (2005)
5. Li, X., Medal, H., Qu, X.: Connected heterogeneous infrastructure location design under additive service utilities. Transp. Res. Part B Methodol. **120**, 99–124 (2019)
6. Neilson, A., Indratmo, Daniel, B., Tjandra, S.: Systematic review of the literature on big data in the transportation domain: concepts and applications. Big Data Res. **17**, 35–44 (2019)
7. Malekzadeh, A., Chung, E.: A review of transit accessibility models: challenges in developing transit accessibility models. Int. J. Sustain. Transp. **14**(10), 733–748 (2020)
8. Stępniak, M., Pritchard, J.P., Geurs, K.T., Goliszek, S.: The impact of temporal resolution on public transport accessibility measurement: review and case study in Poland. J. Transp. Geogr. **75**, 8–24 (2019)
9. Zhou, J., Yang, Y.: Transit-based accessibility and urban development: an exploratory study of Shenzhen based on big and/or open data. Cities **110**(3), 102990 (2021)
10. Harrison, G., Grant-Muller, S., Hodgson, F.: New and emerging data forms in transportation planning and policy: opportunities and challenges for "track and trace" data. Transp. Res. Part C Emerg. Technol. **117**, 102672 (2020)
11. Ana, I., Javier, D., Ibai, L., Maitena, I., Sergio, C.: Big data for transportation and mobility: recent advances, trends and challenges. IET Intell. Transp. Syst. **12**(8), 742–755 (2018)
12. Geurs, K.T., van Wee, B.: Accessibility evaluation of land-use and transport strategies: review and research directions. J. Transp. Geogr. **12**(2), 127–140 (2004)

13. Wang, Y., Shan, X., Meng, Y.: Inter-regional accessibility evaluation model of urban based on taxi GPS big data. Comput. Sci. **46**(01), 271–277 (2019)
14. Jiang, S., Guan, W., He, Z., Yang, L.: Measuring taxi accessibility using grid-based method with trajectory data. Sustainability **10**(9), 3187 (2018)
15. Cui, J., Liu, F., Janssens, D., An, S., Wets, G., Cools, M.: Detecting urban road network accessibility problems using taxi GPS data. J. Transp. Geogr. **51**, 147–157 (2016)
16. Xu, Z., Wei, T., Easa, S., Zhao, X., Qu, X.: Modeling relationship between truck fuel consumption and driving behavior using data from internet of vehicles. Comput.-Aided Civ. Infrastruct. Eng. **33**(3), 209–219 (2018)
17. Yu, W., Zhang, K., Li, J., Sun, H., Qu, Y.: Urban public transport network accessibility based travel data. J. Transp. Syst. Eng. Inf. Technol. **20**(04), 106–112 (2020)
18. Wu, Y., Yang, M., Chen, Z., Bai, S.: Research on bike+ride connection feature mining and site accessibility under multi-source data. In: 2019 China Urban Transportation Planning Annual Conference, pp. 3550–3564. Urban Transportation Planning Academic Committee of China Urban Planning Society, Chengdu (2019)
19. Chen, J., Ni, J., Xi, C., Li, S., Wang, J.: Determining intra-urban spatial accessibility disparities in multimodal public transport networks. J. Transp. Geogr. **65**, 123–133 (2017)
20. Guan, J., Zhang, K., Shen, Q., He, Y.: Dynamic modal accessibility gap: measurement and application using travel routes data. Transp. Res. Part D Transp. Environ. **81**, 102272 (2020)
21. Gao, K., Yang, Y., Sun, L., Qu, X.: Revealing psychological inertia in mode shift behavior and its quantitative influences on commuting trips. Transp. Res. Part F Traffic Psychol. Behav. **71**, 272–287 (2020)
22. Pedro, G., Miguel, P., María, H., Javier, G.: Exploring the potential of mobile phone records and online route planners for dynamic accessibility analysis. Transp. Res. Part A Policy Pract. **125**, 294–307 (2019)
23. Geneviève, B., Ahmed, E.: Daily fluctuations in transit and job availability: a comparative assessment of time-sensitive accessibility measures. J. Transp. Geogr. **52**, 73–81 (2016)
24. Borja, M., Juan, C.: The impacts of congestion on automobile accessibility. What happens in large European cities? J. Transp. Geogr. **62**, 148–159 (2017)
25. Yang, W., Chen, B., Cao, X., Li, T., Li, P.: The spatial characteristics and influencing factors of modal accessibility gaps: a case study for Guangzhou, China. J. Transp. Geogr. **60**, 21–32 (2017)
26. Fayyaz, S., Liu, X., Zhang, G.: An efficient general transit feed specification (GTFS) enabled algorithm for dynamic transit accessibility analysis. PLoS ONE **12**(10), e0185333 (2017)
27. Olle, J., Henrikki, T., Maria, S., Rein, A., Tuuli, T.: Dynamic cities: location-based accessibility modelling as a function of time. Appl. Geogr. **95**, 101–110 (2018)
28. Lee, J., Miller, H.: Analyzing collective accessibility using average space-time prisms. Transp. Res. Part D Transp. Environ. **69**, 205–264 (2019)
29. Wang, L., Cao, X., Li, T., Gao, X.: Accessibility comparison and spatial differentiation of Xi'an scenic spots with different modes based on Baidu real-time travel. Chin. Geogr. Sci. **29**(05), 848–860 (2019)
30. Yan, Y., Guo, T., Wang, D.: Dynamic accessibility analysis of urban road-to-freeway interchanges based on navigation map paths. Sustainability **13**(1), 372 (2021)

Modeling Commercial Vehicle Drivers' Acceptance of Forward Collision Warning System

Yueru Xu, Zhirui Ye, Chao Wang, and Kun Gao

Abstract With the development of computer science, Forward Collision Warning (FCW) systems have been installed in various vehicles in order to improve road safety. Previous studies have been conducted to evaluate the acceptance of FCW systems and explore the contributing factors affecting drivers' attitudes. However, few research studies have focused on the attitudes of commercial vehicle drivers, though commercial vehicle accidents were proved to be more severe than passenger vehicles. This paper tries to examine the attitudes of commercial vehicle drivers toward FCW systems and identify the contributing factors by using a random forests algorithm. FCW data of 24 commercial vehicles were recorded from November 1st to December 21st, 2018 in Jiangsu province. The acceptance rate (FCW records with response) of commercial vehicle drivers for FCW systems is 69.52%. (Acceptance was measured by identifying drivers who reduced their speed in response to a warning from the FCW system.) The accuracy of random forests model is 0.816 after tuning the parameter. In addition, the most important influence variable in this model is vehicle speed with an importance of 0.37. Duration time and warning hour also have significant influence on driver reaction, with values of 0.20 and 0.17, respectively. The results showed that commercial vehicle drivers' acceptance of an FCW system decreases with the increase of vehicle speed. The response time for most cases is timely, usually within 2 s. And the response percentage is higher during daytime than at night. These regularities may be attributable to the larger size and heavier weight of commercial vehicles. The results of this study can help researchers to better understand the behavior of commercial vehicle drivers and to develop more effective FCW systems for commercial vehicles.

Y. Xu (✉)
Intelligent Transportation System Research Center, Southeast University, Nanjing 211189, China
e-mail: xuyr1992@163.com

Z. Ye · C. Wang
School of Transportation, Southeast University, Nanjing 211189, China

K. Gao
Architecture and Civil Engineering, Chalmers University of Technology, 412 96 Goteborg, Sweden

Keywords Commercial vehicles · Forward collision warning system · Driving behavior · Random forests

1 Introduction

Transportation, as a complex system, has caused a lot of concern in many aspects including transportation efficiency [1–3], environmental protection [4–6], and road safety [7, 8]. Among all these aspects, road safety is an extremely important topic. Road accidents cause serious loss of life and property. Among all accidents, those caused by commercial vehicles were more severe and have more consequences than those involving passenger vehicles. In addition, rear-end collisions were the most common crash type in 2015 according to the National Highway Traffic Safety Administration data [9], which accounted for about one third of the total crashes and resulted in 2203 fatalities. In order to prevent this type of crash, a Forward Collision Warning (FCW) system has been installed on more and more vehicles. It can calculate the distance, location and relative speed between the system-equipped vehicle and the vehicle in front of it by monitoring the front vehicle continuously. A visual, audible, and/or tactile signal will occur to alert the driver when the system-equipped vehicle comes too close to the front vehicle. It has significant effects on improving road safety and preventing a forward collision. Up to 70% of rear-end collisions and 20% of all police-reported crashes can be potentially prevented if an FCW system is installed [10].

The FCW system was first proposed by Mercedes-Benz in model year 2000, and then applied on several types of vehicles, including Acura, Audi, and Volvo in the following few years [11]. In the beginning, this function was offered as an optional function in some luxury vehicles. Soon it became more and more common in new model vehicles due to its effectiveness. For most traditional FCW systems, Kalman filtering was the most common algorithm for front vehicle recognition and tracking [12, 13]. With the development of computer science and deep learning, computer vision has become widely used in the development of FCW systems. It can provide more accurate identification results [14–16]. Several FCW algorithms including perceptual-based warning algorithms, Kinematic-based warning algorithms, and others were proposed by researchers to attain a better warning timing [17, 18].

Since the FCW system was proposed, many researchers have explored the effectiveness of this system. Ben-Yaacov et al. [19] evaluated the effect of an imperfect collision avoidance warning system (CAWS) both on driver headway maintenance and the response to warning errors. The result showed that drivers were able to maintain longer and safer time headway after a short adaption of this system. An experiment including 43 drivers for 6 weeks (3 weeks without an FCW system and 3 weeks with an FCW system) were conducted by Shinar and Schechtman [20]. They found that the time spent in short headways (<0.8 s) was reduced by approximately 25% and the time spent in safer headways (>1.2 s) was increased by nearly 20%.

In recent years, Adell et al. [21] evaluated a driver assistance system with 20 test drivers in real traffic conditions along a 50 km long test road. The result indicated that the system can bring faster reaction time and longer time headway, but it may cause slight degeneration of driving skills. Liu et al. [22] carried out more than 500 tests with 55 drivers on a test road in order to understand the effect of a collision avoidance system (CAS). Based on the result, the system can change drivers' behavior and reduce the frequency of dangerous situations if the warning message is timely and accurate. Furthermore, a smart driving system was evaluated by Birrell et al. [23] in real-world driving trials. Two conditions with or without the system were adopted. Key findings from this research showed that the average time headway increased from 1.5 to 2.3 s after the installation of the system.

In addition to evaluating the effectiveness of FCW systems, some researchers focused on the factors influencing drivers' acceptance of this system. Vahidi and Eskandarian [24] conducted a review of FCW systems and concluded that the warning timing of FCW systems has a significant influence on driver acceptance. One study [25] examined an enhanced FCW system with auto braking with a driving simulator study. The analysis of the gaze behavior indicated that driving with the system didn't lead to a stronger involvement in secondary tasks. On the other hand, a negative adaptation was reported by Wege et al. [26] in a study from 30 Volvo trucks driving for approximately 40,000 h for four million kilometers due to an "eyes-off-road effect" during the warning period caused by visual warning signals, which may cause dangerous situations. Li et al. [27] evaluated the driver acceptance of advanced driver assistance systems (ADAS) in Chinese road conditions. They found that there are significant differences in driver acceptance with the variation of driver gender and driver age.

To improve the effect and acceptance of FCW systems, some researchers are trying to develop more intelligent systems by considering drivers' attitudes and reactions. There are mainly two types of intelligent FCW system algorithms. One type combines several risk models in one system and chooses the most suitable model based on driver real-time behavior [28]. James et al. [29] applied this method and proposed an FCW model based on driver acceptance. This model can choose the best risk model based on drivers' brake behaviors. Another type of intelligent FCW system uses one model with a continuous adjustable parameter. This type of system can adjust the parameter values by autonomic learning [30, 31]. Wang et al. [32] proposed an FCW system based on driver behavior and driver characteristics. Based on the result, this model can match driver characteristics throughout a long driving period and change the model parameter in a timely manner. Pei et al. [33] proposed a new parameter Tbuffer to establish its warning strategy, which is a modification of time to collision (TTC) by taking both vehicles into consideration.

We find that many efforts have been conducted to evaluate the influence factors of driver acceptance on FCW systems and improve system efficiency. However, most FCW systems were designed based on private cars. Relevant studies on the influence factors of commercial vehicle drivers' acceptance are rare. Therefore, it's essential for researchers to evaluate the influence factors of commercial vehicle driver acceptance on FCW systems to help better design the system for commercial vehicles.

Therefore, this article will focus on the influence factors of commercial driver acceptance of FCW systems. A random forests model was applied. Driver reaction was used as a dependent variable and several driver characteristics and road characteristics were used as input variables. The results of this study can help system designers better understand the influence mechanism of FCW systems on commercial vehicles and help to design a differentiated FCW system for commercial vehicles based on various factors.

2 Data Extraction

The data that was used in this study is from the intelligent data platform of commercial vehicles in Jiangsu province. An FCW system (designed by Mobileye) was installed in all these commercial vehicles. The study gathered the warning information from 24 vehicles (12 buses and 12 trucks) from November 1st to December 21st in 2018. Since the majority of commercial vehicle drivers are male, the 24 drivers are all male drivers, including 9 young drivers (18–35 years old), 11 mid-aged drivers (35–50 years old), and 4 older drivers (more than 50 years old).

Using the data systems of commercial vehicles in Jiangsu province, all the FCW signals were recorded and uploaded to the platform. Each record contains several variables including driver ID, warning start time, warning end time, vehicle speed before warning, vehicle speed after warning, vehicle GPS. 1706 FCW records were collected during 51 days. First, we preprocessed these records to obtain more useful information.

The GPS data of each record was transferred to a Baidu coordinate system. The location of each record was then identified. In this study, we separated FCW records into two different road types, urban road and highway.

The driver acceptance of each record is measured by acceleration. We divided acceleration into response and no response. The driver is regarded as showing a "response" if the vehicle speed before FCW signals is larger than the speed after FCW signals, otherwise, the driver is regarded as "no response."

3 Methodology

In this study, a random forests algorithm was applied. It is one of the most common machine learning methods and was proposed by Breiman in 2001 [34]. This method consists of numerous decision trees. All these decision trees are independent. The result of random forests can be acquired by combining the results of all decision trees. This method can improve the accuracy substantially without adding a lot of computation. It's more suitable than other machine learning methods in classification problems. In addition, it can obtain the importance of influence factors by using out of bag (OOB) data.

3.1 Decision Trees

Before introducing random forests, a decision tree, the sub-unit of random forests should be introduced. A decision tree can also be called a classification tree or regression tree, depending on the output variable type (classified variable or continuous variable). This method was first proposed by Hunt et al. [35] in 1966. It can be expressed as a tree-like structure. Each internal node in the tree represents a "test" on an attribute, each branch represents the different result of the test, and each leaf node represents a class label. The process from root to leaf represents classification rules. A binary tree example is shown in Fig. 1.

For the process of decision tree establishment, the most important part is decision rules [36]. The rules can help to derive the best criterion of internal nodes and choose the best split point. In this study, we use the Gini index to measure the purity of data.

The Gini index is a common method for data purity measurement. It can also be used to divide the characteristics. The expression of a Gini index is shown as Eq. 1.

$$Gini(D) = 1 - \sum_{i=1}^{m} p_i^2 \qquad (1)$$

Gini(D) means randomly select 2 samples from dataset D, the probability that the categories of two samples are inconsistent. p_i means the probability of category i in the whole dataset D. If we choose condition A as a classification standard, the Gini index after the split can be expressed as follows:

$$Gini_A(D) = \sum_{j=1}^{n} \frac{|D_j|}{|D|} \times Gini(D_j) \qquad (2)$$

Therefore, the Gini index gain value can be calculated as:

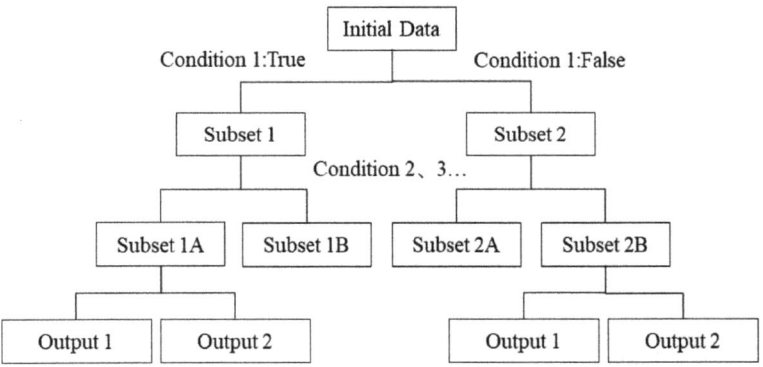

Fig. 1 Structure of a binary tree

$$\Delta Gini(A) = Gini(D) - Gini_A(D) \tag{3}$$

When selecting the best criterion of internal nodes, we should choose the one with the largest $\Delta Gini(A)$.

3.2 Random Forests

From the introduction above, we know that a decision tree can handle classification problems effectively. However, in most cases, a single decision tree is inadequate for a good result. Therefore, the concept of ensemble was applied. In this concept, to solve the inherent shortcomings of a single model, we combine several models together and obtain the results by considering the results of all these models. The random forests algorithm is the outcome of this concept.

The main steps of the random forests algorithm can be summarized as follows:

Step 1: For each decision tree, the training set is drawn by sampling with replacement, which is called the bagging method. This is one characteristic of random forests. About one-third of the cases are left out of the samples. These cases are called out-of-bag (OOB) data [37]. The data can be used in the following steps to get an error estimate and evaluate the importance of variables.

Step 2: After all the samples are selected, the input and output variables should be run down the tree. Each tree is grown to the maximum depth. At the end of the run, the result of each decision tree can be drawn. The result of the random forests algorithm is obtained by averaging all the results of the individual decision trees. Since there are a large number of decision trees, the generalization error is limited. In other words, an over-fitting problem is impossible in the random forests algorithm.

Step 3: The OOB error is estimated after getting the result. In random forests, there is no need for cross-validation or a separate test set to get an unbiased estimate of the result [38]. It has been estimated internally during the run. The OOB error of the prediction can be calculated by:

$$ER = n^{-1} \sum_{i=1}^{n} I\left(Y^{OOB}(X_i) \neq Y_i\right) \tag{4}$$

where I(*) is the indicator function; n means the number of trees in the random forests; Xi means the input of ith OOB data; Yi means the actual output of ith OOB data; and YOOB(Xi) represents the predicted output calculated by random forests.

Step 4: The variable importance should be calculated to understand the importance of each input variables on the results. In random forests, two different kinds of methods are mainly used to calculate the variable importance: Permutation importance and GINI importance. The accuracy of permutation importance is a little higher than GINI importance if the input variables have both classified variables and continuous variables. Therefore, we choose permutation importance in this article. It's also based on OOB data. We use VIM_{OOB} to represent the permutation importance of the variable. It can be calculated by:

$$VIM_{OOB}(ij) = \frac{\sum_{p=1}^{n_o^j} F\left(Y_p = Y_{p,X_i}^j\right)}{n_o^j} \tag{5}$$

In which F(*) is the indicator function; n_o^j means the number of OOB data in jth decision tree. Y_p is the actual output of pth OOB data; Y_{p,X_i}^j means the predicted result after Xi is permuted.

Finally, the permutation importance can be calculated by:

$$VIM_{OOB}(i) = \frac{1}{n} \sum_{j=1}^{n} VIM_{OOB}(ij) \tag{6}$$

In this study, the Python software was used to develop the random forests algorithm and calculate the OOB error and importance.

4 Results and Analysis

A total of 1706 valid FCW records were identified during 51 days. The statistics of FCW record data characteristics are shown in Table 1. In this table, vehicle speed means commercial vehicle speed at the start time of FCW signals. Warning hour

Table 1 Statistics of FCW record data characteristics

Variable	Maximum	Minimum	Average	SD
Vehicle speed	100	30	60.57	19.76
Warning hour	23	0	13.00	3.97
Duration time	5	1	3.05	1.19
Variable	Type		Frequency	
Road type	1 (Urban road)		1407	
	2 (Highway)		299	
Driver age group	1 (Young)		670	
	2 (Mid-aged)		518	
	3 (Elder)		518	
Vehicle type	1 (Passenger car)		1066	
	2 (Truck)		640	
Driver reaction	0 (No response)		520	
	1 (Response)		1186	

means the time of t day (0 a.m.–23 p.m.) when an alarm occurs, from 1 o'clock to 23 o'clock. The duration time means the time interval between warning start time and warning end time. In this study, the maximum duration time is 5 s.

Several classified variables including road type, driver age group, and vehicle type were applied as inputs in this study. Different numbers are used to represent different types of variables. These details are also listed in Table 1.

For the output driver reaction, 0 means no response and 1 means response. We found 1186 response cases among 1706 samples. The acceptance percentage of commercial vehicle drivers is 69.52%. Therefore, the driver acceptance rate was not as high as expected. The influence of various variables on driver acceptance should be explored to better design the system and improve the acceptance.

The random forests analysis was conducted using the binary output variable, response or no response. Input variables included vehicle speed, time, duration time, road type, driver age group, and vehicle type. In random forests, two parameters should be tuned in order to obtain an optimal performance. One parameter is called m_try, which means the number of input variables in each decision tree. The other one is called n_tree, which means the number of decision trees in a random forests algorithm.

For the parameter m_try, the common value selected is $\sqrt{N}$, where N represents the total number of input variables. In this case, a total of 6 variables were applied, so the value of m_try can be 2 or 3. In addition, n_tree is usually identified as a value that provides stable OOB errors. It can be confirmed by comparing the OOB error of different parameters from low to high. Therefore, we drew a line graph to represent the accuracy tendency (1-ER) when n_tree increases. Both m_try = 2 and m_try = 3 were considered when drawing this figure.

In Fig. 2, we see that at the very start, the accuracy increases with the increase of n_tree and then becomes stable. When m_try = 2 and n_tree = 194, the accuracy

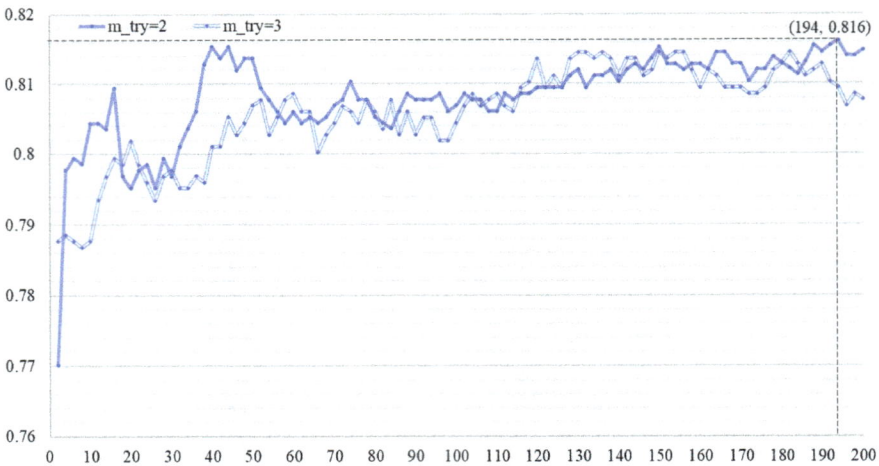

Fig. 2 OOB error tendency under different parameters

Fig. 3 Normalized variable permutation importance

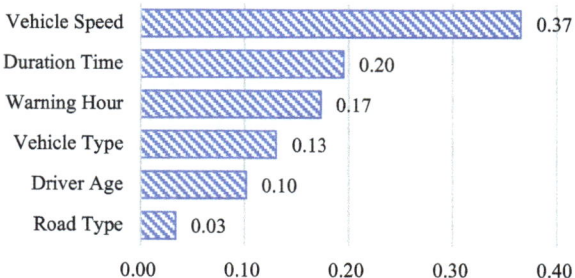

reached the largest value. Therefore, the most suitable parameter values in this case are m_try = 2 and n_tree = 194. In this situation, the accuracy is 0.816. It's relatively high and the results of the model can be used for further analysis.

The permutation importance of each variable was calculated when running the random forests model. After normalization, the importance for each factor was displayed in Fig. 3. Among the 6 influence factors, we find that the most important variable affecting commercial vehicle driver reaction is vehicle speed. The importance value is 0.37. Duration time and warning hour also have significant influence on driver reaction. The values are 0.20 and 0.17, respectively. Road type has the lowest importance among the 6 variables—the importance is only 0.03.

Based on the importance value, we choose three key variables, vehicle speed, duration time, and warning hour to further analyze the potential characteristics of driver reaction and relevant influence factors. The importance values of these three variables are all larger than 0.15.

4.1 Vehicle Speed

Figure 4 represents the relationship between vehicle speed and acceleration after the FCW warning signals. Vehicle speed is the speed at the start time of FCW signals and acceleration is the average acceleration between the start time of FCW signals and the end time of FCW signals. Drivers' reaction is more obvious if the absolute value of acceleration is larger.

From this figure, we find that most drivers have an obvious reaction to FCW signals while the initial vehicle speed is low (30–60 km/h). With the increase of speed, the percentage of obvious reaction reduced. When vehicle speed is larger than 60 km/h, the acceleration value becomes small (less than −2.00 m²/s).

Therefore, we can conclude that the commercial vehicle driver acceptance of FCW system decreases with the increase of vehicle speed. It may be attributable to the fact that commercial vehicles have greater size and weight than private cars and are harder to decelerate. Drivers prefer to maintain the speed and don't brake hard when the original speed is high.

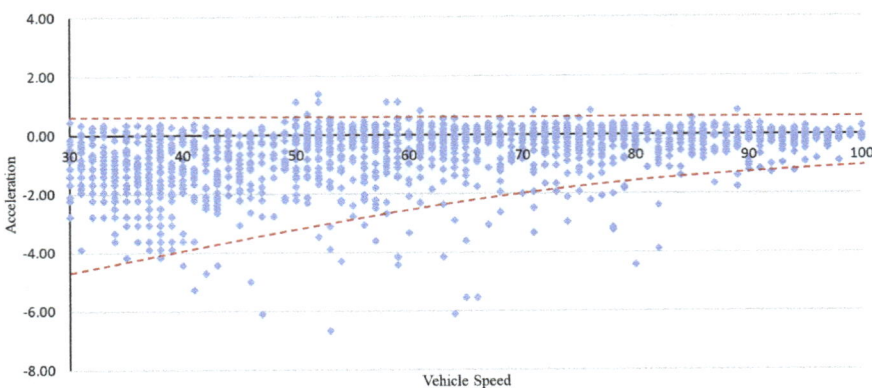

Fig. 4 Relationship between vehicle speed and acceleration

4.2 Duration Time

The number of FCW records for the two types of responses were counted based on different duration time (from 1 to 5 s) and shown in a heatmap (Fig. 5). In this heatmap, a darker color means a higher percentage. We can conclude that when drivers respond to FCW signals, the duration times are mainly 1 s (28.83%) or 2 s (31.45). When duration time increases, the percentage decreases. When drivers have no response to FCW signals, the duration times are usually 3 and 4 s. This suggests that these drivers didn't take any steps to respond after 3–4 s warning.

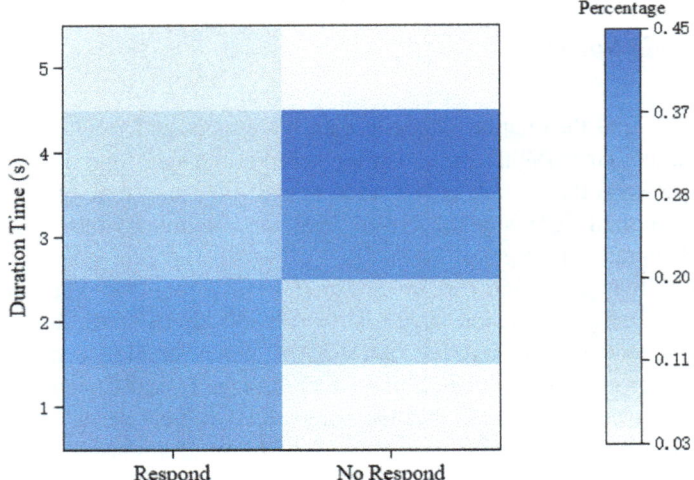

Fig. 5 Heatmap of duration time

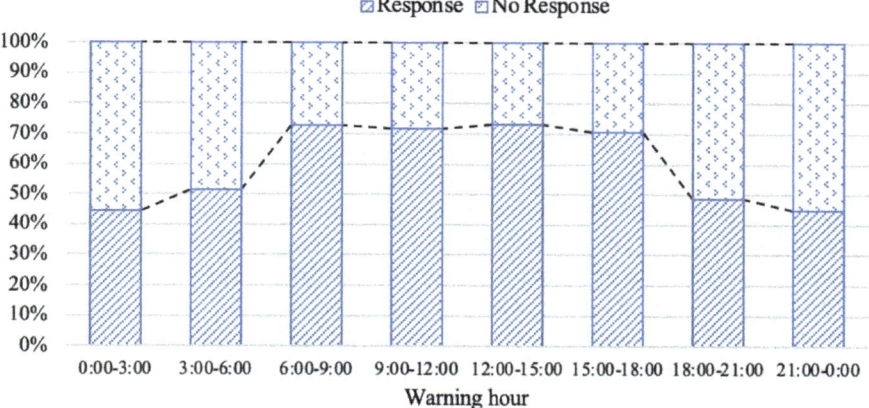

Fig. 6 Histogram of warning hour

Therefore, it can be concluded that commercial drivers prefer to respond to FCW signals right away, mainly in less than 2 s. If they choose to ignore the warning, longer warning signals won't have a significant effect.

4.3 Warning Hour

The histogram of warning hour is shown in Fig. 6. In this figure, we divided one day into 8 time periods. Every time period contains 3 h. The percentage of response and no response in each time period was calculated and drawn as a bar.

In this figure, we can find that between 6:00 and 18:00, commercial vehicle drivers have higher response percentage on FCW signals. The percentages in these time periods are all larger than 70%. Furthermore, it can be easily observed that this time period (6:00–18:00) mainly falls within daytime.

In other words, we find that commercial vehicle drivers have higher response percentages on FCW signals in the daytime than at night. This result is contrary to what we expected. When the light condition is good, these drivers prefer to trust the system and respond to FCW signals. When the day becomes dark, drivers prefer to trust their own judgment and ignore FCW signals. This phenomenon may also be due to the larger size and weight of commercial vehicles.

In general, vehicle speed, duration time, and warning hour are three key influence factors affecting commercial drivers' attitudes towards FCW systems. Drivers have higher acceptance of FCW systems during daytime when commercial vehicles' speeds are relatively low. The acceptance may decrease at night when commercial vehicles' speeds are relatively high. Driver responses to FCW systems are sensitive and timely, usually within 2 s.

5 Conclusion

The investigation of drivers' attitudes towards FCW systems is very important in human factor research. Previous studies have been applied to evaluate the acceptance of FCW systems and identify the contributing factors affecting drivers' attitudes. However, most studies have focused on private cars, and few have addressed commercial drivers' attitudes toward FCW systems. Therefore, this paper aimed to examine the attitudes of commercial vehicle drivers toward FCW systems under the influence of several factors by developing a random forests model.

The data of 24 commercial vehicles including 12 buses and 12 trucks from the data systems of commercial vehicles in Jiangsu province were used in this study. A random forests algorithm was applied i to model the relationship between driver response and relevant variables. After tuning the parameters, the accuracy is 0.816, which is relatively high for further analysis. The permutation importance of each variable was calculated. The most important influence variable is vehicle speed with an importance of 0.37, followed by duration time and warning hour, with an importance of 0.20 and 0.17, respectively.

In addition, we found that commercial vehicle drivers' acceptance of an FCW system decreases with the increase of vehicle speed. They prefer to respond to FCW signals in a timely manner, usually in less than 2 s. The response percentage is higher during daytime than at night. These regularities may be due to the fact that commercial vehicles have larger size and heavier weight. Therefore, it may be beneficial to increase the sensitivity of FCW systems during daytime when vehicle speed is low and improve the accuracy of FCW systems at night when vehicle speed is high. This may improve the acceptance of the system by commercial vehicle drivers. The alarm form for FCW warnings can be short and rapid signals since the responses to FCW signals are timely.

This paper explored the influence factors of commercial drivers' attitudes towards FCW systems. The results of this study can help system designers better understand the behavior of commercial vehicle drivers and develop more effective FCW systems for commercial vehicles.

In this study, the accuracy of FCW system is regarded as relatively high and be consistent at all times since Mobileye is a leading company of ADAS systems and have high quality of products. But we also plan to discuss the accuracy of FCW system under different situations and investigate commercial vehicle drivers about the system accuracy in the future to further understand the acceptance.

References

1. Gao, K., Yang, Y., Sun, L., Qu, X.: Revealing psychological inertia in mode shift behavior and its quantitative influences on commuting trips. Transp. Res. Part F Traffic Psychol. Behav. **71**, 272–287 (2020)

2. Wu, J., Kulcsár, B., Ahn, S., Qu, X.: Emergency vehicle lane pre-clearing: from microscopic cooperation to routing decision making. Transp. Res. Part B Methodol. **141**, 223–239 (2020)
3. Li, X., Medal, H., Qu, X.: Connected infrastructure location design under additive service utilities. Transp. Res. Part B Methodol. **120**, 99–124 (2019)
4. Gao, K., Yang, Y., Li, A., Li, J., Yu, B.: Quantifying economic benefits from free-floating bike-sharing systems: a trip-level inference approach and city-scale analysis. Transp. Res. Part A Policy Pract. **144**, 89–103 (2021)
5. Xu, Y., Zheng, Y., Yang, Y.: On the movement simulations of electric vehicles: a behavioral model-based approach. Appl. Energy **283**, 116356 (2021)
6. Qu, X., Yu, Y., Zhou, M., Lin, C.T., Wang, X.: Jointly dampening traffic oscillations and improving energy consumption with electric, connected and automated vehicles: a reinforcement learning based approach. Appl. Energy **257**, 114030 (2020)
7. Meng, Q., Qu, X.: Estimation of rear-end vehicle crash frequencies in urban road tunnels. Accid. Anal. Prev. **48**, 254–263 (2012)
8. Kuang, Y., Qu, X., Wang, S.: A tree-structured crash surrogate measure for freeways. Accid. Anal. Prev. **77**, 137–148 (2015)
9. National Highway Traffic Safety Administration: Traffic safety facts, 2012 data: pedestrians. Ann. Emerg. Med. **65**(4), 452 (2015)
10. Jermakian, J.S.: Crash avoidance potential of four passenger vehicle technologies. Accid. Anal. Prev. **43**(3), 732–740 (2011)
11. Cicchino, J.B.: Effectiveness of forward collision warning and autonomous emergency braking systems in reducing front-to-rear crash rates. Accid. Anal. Prev. **99**, 142–152 (2017)
12. Kiefer, R.J., LeBlanc, D., Palmer, M.D., Salinger, J., Deering, R.K., Shulman, M.: Development and validation of functional definitions and evaluation procedures for collision warning/avoidance systems (No. DOT-HS-808-964). United States. Department of Transportation. National Highway Traffic Safety Administration (1999)
13. Zhang, D., Li, K., Wang, J.: Radar-based target identification and tracking on a curved road. Proc. Inst. Mech. Eng. Part D J. Automob. Eng. **226**(1), 39–47 (2012)
14. Lei, Y., Wu, J.: Study of applying ZigBee technology into foward collision warning system (FCWS) under low-speed circumstance. In: 2016 25th Wireless and Optical Communication Conference (WOCC), May 2016, pp. 1–4. IEEE
15. Lewis, B.A., Eisert, J.L., Baldwin, C.L.: Validation of essential acoustic parameters for highly urgent in-vehicle collision warnings. Hum. Factors **60**(2), 248–261 (2018)
16. Iranmanesh, S.M., Mahjoub, H.N., Kazemi, H., Fallah, Y.P.: An adaptive forward collision warning framework design based on driver distraction. IEEE Trans. Intell. Transp. Syst. **19**(12), 3925–3934 (2018)
17. Wang, X., Chen, M., Zhu, M., Tremont, P.: Development of a kinematic-based forward collision warning algorithm using an advanced driving simulator. IEEE Trans. Intell. Transp. Syst. **17**(9), 2583–2591 (2016)
18. Elmalaki, S., Tsai, H.R., Srivastava, M.: Sentio: driver-in-the-loop forward collision warning using multisample reinforcement learning. In Proceedings of the 16th ACM Conference on Embedded Networked Sensor Systems, Nov 2018, pp. 28–40. ACM
19. Ben-Yaacov, A., Maltz, M., Shinar, D.: Effects of an in-vehicle collision avoidance warning system on short- and long-term driving performance. Hum. Factors **44**(2), 335–342 (2002)
20. Shinar, D., Schechtman, E.: Headway feedback improves intervehicular distance: a field study. Hum. Factors **44**(3), 474–481 (2002)
21. Adell, E., Várhelyi, A., Dalla Fontana, M.: The effects of a driver assistance system for safe speed and safe distance—a real-life field study. Transp. Res. Part C Emerg. Technol. **19**(1), 145–155 (2011)
22. Liu, H., Wei, H., Yao, Z., Ai, Q., Ren, H.: Effects of collision avoidance system on driving patterns in curve road conflicts. Procedia Soc. Behav. Sci. **96**, 2945–2952 (2013)
23. Birrell, S.A., Fowkes, M., Jennings, P.A.: Effect of using an in-vehicle smart driving aid on real-world driver performance. IEEE Trans. Intell. Transp. Syst. **15**(4), 1801–1810 (2014)

24. Vahidi, A., Eskandarian, A.: Research advances in intelligent collision avoidance and adaptive cruise control. IEEE Trans. Intell. Transp. Syst. **4**(3), 143–153 (2003)
25. Muhrer, E., Reinprecht, K., Vollrath, M.: Driving with a partially autonomous forward collision warning system: how do drivers react? Hum. Factors **54**(5), 698–708 (2012)
26. Wege, C., Will, S., Victor, T.: Eye movement and brake reactions to real world brake-capacity forward collision warnings—a naturalistic driving study. Accid. Anal. Prev. **58**, 259–270 (2013)
27. Li, G., Li, S.E., Cheng, B.: Field operational test of advanced driver assistance systems in typical Chinese road conditions: the influence of driver gender, age and aggression. Int. J. Automot. Technol. **16**(5), 739–750 (2015)
28. Hirose, T., Oguchi, Y., Sawada, T.: Framework of tailormade driving support systems and neural network driver model. IATSS Res. **28**(1), 108–114 (2004)
29. James, D.J.G., Boehringer, F., Burnham, K.J., Copp, D.G.: Adaptive driver model using a neural network. Artif. Life Robot. **7**(4), 170–176 (2004)
30. Rajaonah, B., Tricot, N., Anceaux, F., Millot, P.: The role of intervening variables in driver–ACC cooperation. Int. J. Hum. Comput. Stud. **66**(3), 185–197 (2008)
31. Chang, C.Y., Chou, Y.R.: Development of fuzzy-based bus rear-end collision warning thresholds using a driving simulator. IEEE Trans. Intell. Transp. Syst. **10**(2), 360–365 (2009)
32. Wang, J., Yu, C., Li, S.E., Wang, L.: A forward collision warning algorithm with adaptation to driver behaviors. IEEE Trans. Intell. Transp. Syst. **17**(4), 1157–1167 (2015)
33. Pei, X.F., Qi, Z.Q., Wang, B.F., Liu, Z.D.: Vehicle frontal collision warning/avoidance strategy. J. Jilin Univ. (Eng. Technol. Ed.) **44**(3), 599–604 (2014)
34. Breiman, L.: Random forests. Mach. Learn. **45**(1), 5–32 (2001)
35. Hunt, E.B., Marin, J., Stone, P.J.: Experiments in induction (1966)
36. Quinlan, J.R.: Simplifying decision trees. Int. J. Man Mach. Stud. **27**(3), 221–234 (1987)
37. Lunetta, K.L., Hayward, L.B., Segal, J., Van Eerdewegh, P.: Screening large-scale association study data: exploiting interactions using random forests. BMC Genet. **5**(1), 32 (2004)
38. Breiman, L.: Bagging predictors. Mach. Learn. **24**(2), 123–140 (1996)

Author Index